SCIENTIFIC HUMANISM

The Works of Lothrop Stoddard

Volume eleven

ISBN: 978-1-59364-070-5 Softcover

ISBN: 978-1-59364-071-2 Hardcover

Liberty Bell Publications
655 Sandifer Road
York, SC 29745
1-803-818-5407
dmartyo@protonmail.com

Liberty Bell Publications® is a registered trademark of Liberty Bell Publications LLC, a South Carolina Limited Liability Company.

Printed in the United States of America

EDITOR'S NOTE

Those who cannot remember the past are condemned to repeat it.

—George Santayana, *The Life of Reason* (1905)

LIBERTY BELL PUBLICATIONS IS REPRINTING this controversial book from 1926 as the eleventh volume in a new series entitled *The Works of Lothrop Stoddard.*

It is amazing how urgent the matters discussed here are, even more so today than they were 100 years ago.

What a paradox modern life is! On the one hand, we see a sudden extension of human knowledge and power so stupendous as to stagger the imagination. Within the past century, the slow, slow tempo of material progress by which, through countless ages, man climbed painfully from savagery to civilization, has been violently shattered, and we have literally leaped into a new self-made world. The hidden powers of nature have yielded themselves wholesale, as though at the wave of a magician's wand. Our dwelling place, the earth, shrinks to the measure of human hands. Cosmic wonders, from the fiery whorls of outer nebula to the pulsing heart of the minute atom, come within our ken. Most significant of all, even the mysterious depths of our own selves are being plumbed, and the hope of ancient sages, *"Know Thyself!"* seems on the way to realization.

Such is the brilliant side of our age. But there is another—and a darker—side. This new world, which we have suddenly entered, resplendent though it be, proves likewise in many respects a troublesome and perplexing abode. Unprepared by previous experience, swayed by ideas and habits brought with us from other times, and confused by the very magnitude of our new powers and the ill-digested mass of our new knowledge, we are like pioneers in a strange land, perplexedly feeling our way amid unfamiliar scenes.

Indeed, it is not yet certain that we can permanently dwell in this new world of ours. The deeper we probe the recesses of human nature, the more painfully do we realize man's innate weaknesses and limitations. It is far from impossible that we may have undertaken a task beyond our ability and that modern civilization may be headed for disaster.

Where possible, we have corrected grammatical errors, including punctuation and obsolete spelling. We have deleted hyphens in words that were hyphenated in 1926 but no longer are. No correction has altered or changed the author's intent or meaning.

We intend to publish the other works of the author as time permits.

Please send any corrections or ideas to the publisher.

Liberty Bell Publications

THE WORKS OF LOTHROP STODDARD

VOLUME 1: REFORGING AMERICA

VOLUME 2: THE STORY OF YOUTH

VOLUME 3: THE RISING TIDE OF COLOR

VOLUME 4: THE REVOLT AGAINST CIVILIZATION

VOLUME 5: SOCIAL CLASSES IN POST-WAR EUROPE

VOLUME 6: CLASHING TIDES OF COLOR

VOLUME 7: LUCK YOUR SILENT PARTNER

VOLUME 8: RACIAL REALITIES IN EUROPE

VOLUME 9: THE FRENCH REVOLUTION IN SAN DOMINGO

VOLUME 10: THE NEW WORLD OF ISLAM

VOLUME 11: SCIENTIFIC HUMANISM

VOLUME 12: STAKES OF THE WAR

VOLUME 13: PRESENT DAY EUROPE

VOLUME 14: MASTER OF MANHATTAN

VOLUME 15: INTO THE DARKNESS

VOLUME 16: THE SHORT WRITINGS

VOLUME 17: LONELY AMERICA

SCIENTIFIC HUMANISM

SCIENTIFIC HUMANISM

BY

LOTHROP STODDARD

A.M., PH.D. (HARV.)

CHARLES SCRIBNER'S SONS

NEW YORK • LONDON

1926

Printed in the United States of America

TO

MY WIFE

FOREWORD

Are we entering a new age? Will we adjust ourselves to the new world that modern science has disclosed? Is there to be a new flowering of the mind and spirit quickened by the vast extensions of knowledge and power with which we have been endowed, as the Renaissance flowered in Humanism four centuries ago? Shall our present transition time, with all its unparalleled possibilities for good and for ill, culminate in a great civilization or in a great catastrophe?

Such are the queries that today stir forward-looking minds. What the outcome will be, no one knows. Yet we may safely predict that the outcome will be largely determined by what we, of this transition epoch, think and do. To offer some suggestions for sound thought and action, this book has been written.

Lothrop Stoddard

Brookline, Massachusetts
February 8th, 1926

CONTENTS

CHAPTER I

OUR SCIENTIFIC AGE

OURS IS AT ONCE A glorious and a perilous age: glorious, because of its prospects of unprecedented good; perilous, because of its possibilities of unparalleled ill. Never before in all its long history has mankind confronted such alternatives of triumph or disaster.

What a sonorous epic, what a heart-kindling saga of the *Ascent of Man* from his humble origins to his present overlordship of the earth some great poet must compose! Consider this eloquent portrayal of man's material progress: "The change in the relation of man to his environment—of his place in nature—is indeed astounding. Its recital is that of a fairy romance. This little hairless animal that once crept, naked and forlorn, over the face of the earth, going warily and timorously amidst its numberless perils, snatching a precarious subsistence from chance scraps picked up by the way, clutching, as its choicest possession, a chunk of stone or a knotted stick; the sport of the elements, the prey of the larger beasts; homeless and shelterless save for the cavern whose tenancy he disputed with the hyena or the cave-bear: behold him now in all the opulence of his great inheritance of knowledge, lording it over the world through which he once slunk in momentary peril of his life! Less gifted with specialized means than any other animal, he has yet outstripped them all. He burrows into the very bowels of the earth; he traverses its surface at a speed that leaves the fleetest beast as stationary; he follows the leviathan into the depths of the sea; he skims the surface of the ocean more swiftly than the flying fish; he soars to aerial heights inaccessible to the eagle or the condor."[1]

What a paradox modern life is! On the one hand, we see a sudden extension of human knowledge and power so stupendous as to stagger the imagination. Within the past century, the slow, slow tempo of material progress by which, through countless ages, man climbed painfully from savagery to civilization, has been violently shattered, and we have literally leaped into a new self-made world. The hidden powers of nature have yielded themselves wholesale, as though at the wave of a magician's wand. Our dwelling place, the earth, shrinks to the measure of human hands. Cosmic wonders, from the fiery whorls of outer nebula to the pulsing heart of the minute atom, come within our ken. Most significant of all, even the mysterious depths of our own selves are being plumbed, and the hope of

1. R. A. Freeman, *Social Decay and Regeneration,* pp. 13-14 (London, 1921).

ancient sages, *"Know Thyself!"* seems on the way to realization.

Such is the brilliant side of our age. But there is another—and a darker—side. This new world, which we have suddenly entered, resplendent though it be, proves likewise in many respects a troublesome and perplexing abode. Unprepared by previous experience, swayed by ideas and habits brought with us from other times, and confused by the very magnitude of our new powers and the ill-digested mass of our new knowledge, we are like pioneers in a strange land, perplexedly feeling our way amid unfamiliar scenes.

Indeed, it is not yet certain that we can permanently dwell in this new world of ours. The deeper we probe the recesses of human nature, the more painfully do we realize man's innate weaknesses and limitations. It is far from impossible that we may have undertaken a task beyond our ability and that modern civilization may be headed for disaster. One well-known philosopher warns us of the fate of Icarus,[2] the fabled youth of Greek mythology, who, rashly venturing to fly, crashed down to earth in speedy ruin. Other observers, yet more somber, go so far as to predict the possible extinction of all mankind.[3]

Whether these warning voices be true or false, one thing is certain: we moderns will never voluntarily quit this new world of ours. Another thing equally certain is that we must adjust ourselves to our new world or suffer the direst penalties, because man, like every other living being, is bound by nature's stern command to adapt or perish.

However, the closer we scrutinize our situation, the more hopeful appear our prospects of thus adjusting ourselves. The truth of the old adage, "Knowledge is Power," was never better exemplified than at the present day. Indeed, it is imperfect appreciation of our new knowledge that is at the root of half our difficulties. We have, already on hand, knowledge enough to solve our most pressing problems, if we have but the wit to discern and the intelligence to apply.

It is this vast extension of knowledge that not only distinguishes our age from past times but which is also our chief hope for the future. Other civilizations perished because they literally did not know what the matter with them was. Our civilization has no such excuse. Today, sound diagnosis of our ills has become possible and is becoming increasingly accurate. In fact, much of our so-called "literature of despair" is not so much pessimism as clinical examination. The very fact that we are so keenly aware of our ailments is the best augury that we are going to find cures for them. The mere presence of danger is, in itself, no reason for nervous dread. Indeed, to the healthy, normal man, the spice of danger is apt to be rather a challenge and an inspiration. Only when peril looms overwhelming and inescapable does despair chill the heart and paralyze the will. And surely no well-informed observer of our times can dogmatically assert that we are in any such hopeless posture.

How, then, shall we characterize this vast extension of human knowledge that has led us into our new world, with all its unprecedented possibilities for good and

2. Bertrand Russell, *Icarus or the Future of Science* (New York, 1924).

3. For typical comments along these lines, see: Freeman, *op. cit.;* S.A. Coblentz, *The Decline of Man* (New York, 1925).

ill? It can be summed up in one word: *Science*. Ours is unquestionably *The Scientific Age*. It is that which distinguishes our epoch from all past times. Our civilization may be inferior to others in certain respects. It may lack the beauty of the Greek, the durability of the Chinese, and the profound mysticism of the Hindu. But in effective knowledge of facts and control over natural forces, our age has achieved triumphs of which past ages did not even dream.

Now, do we realize what this truly means? Do we grasp the full significance of this mighty advance in human progress? Too often the word "science" evokes merely a composite picture of certain mechanical inventions that intimately touch our daily lives. These are not "science." They are mere random fruits of something infinitely vaster in scope and infinitely deeper in meaning. Science is, in its essence, a method, a process, and an attitude of mind. However marvelous those inventions, however rich the store of facts gained from the study of nature, they are but the tools and counters used by the guiding spirit behind. And this guiding spirit, this *Spirit of Science,* is a desire for *truth* so strong and so compelling that prejudices and preconceptions are burned away, leaving the mind crystal clear to perceive the significance of fresh knowledge and adjust it harmoniously to knowledge already acquired. And that, in turn, means a mental attitude both free and flexible, capable of progress by methods at once steady and sure.

Such is the spirit of science—the stimulus of our time and the hope of the future. If it shall come more and more to influence our actions and outlook, we may rest assured of a brighter tomorrow. Specific "facts" and theories now held as firmly established may turn out to be partly, or even wholly, untrue; existing institutions and ideals may have to be utterly revised; our very conception of the universe may prove to be an illusion. Nevertheless, so long as we confront our problems with the serene determination to attain more knowledge, more light, and more truth, we shall emerge from our time of trial not merely on a sounder basis but also having undergone a minimum of suffering and disturbance during the process of transition. There is no basic reason, for instance, why changes in the political or social spheres may not be envisaged in the same way as changes in the field of abstract science, provided the need for such an attitude be sufficiently clear and the will to so act be sufficiently strong. Assuredly the scientific spirit is applicable to every phase of human activity—that spirit of impartial observation, tireless research, unprejudiced deduction, and tolerant cooperation, which today characterizes the true scientist, and which alone has made possible the vast extensions of knowledge that have occurred in recent times. We have but to continue and amplify this spirit and these methods to ensure even greater results than those that have thus far been achieved.

Incalculably great though these results may be, the hopeful thing is that they require no mysterious methods or superhuman qualities to bring them about. Huxley pithily described science as merely "trained and organized common sense," and the methods of science as but extensions of those practiced in everyday life.[4]

4. T. H. Huxley, "On the Educational Value of the Natural History Sciences," in *Collected Essays,* volume entitled: "Science and Education."

The general method of science is thus simply summarized: "to explore the facts of nature, to ascertain their mutual relations, and to arrange them as far as possible into a consistent and intelligible scheme."[5]

The novel feature of modern science lies not in its sphere of activity nor in the mental qualities evoked in its service. "Science is the product of human reason applied to the phenomena of nature. It is, therefore, as old as rational thought. The straight-thinking man was always a scientist."[6] Indeed, the beginnings of scientific interest run far back into the mists of time. Some ancient peoples, like the Egyptians and Babylonians, made considerable progress along scientific lines such as astronomy and engineering, while the Greeks displayed an intellectual grasp and breadth of vision unsurpassed till our own days.

What specifically distinguishes modern science from its forerunners is its unswerving quest for objective truth, attained through exact and verifiable knowledge. This cardinal principle, erected into a clear-cut ideal, consciously held by a large class of persons, and widely diffused through the general population, invests our age with its peculiarly scientific character and marks it off from all previous times. Even the greatest intellects of the past were limited to meager stocks of fact-data and usually mixed their scientific observations with philosophical speculations or religious dogmas, which vitiated their conclusions. Furthermore, the scientists of olden days were either solitary thinkers or isolated groups of scholars always unintelligible to, and often persecuted by, their contemporaries. Under such circumstances no genuine scientific flowering was possible.

Today, conditions are radically changed. Science has emerged from its long obscurity and actually sets the tone of our age. Not only has it transformed our material surroundings, but it is also revolutionizing our thinking and our ideals. By it the human mind is being raised to a higher plane of activity, and we seem to be only at the beginning of a novel line of intellectual development. Here, indeed, is something really new in human history!

Furthermore, there seems to be no basic reason why this triumphant advance should not continue, not merely at its present rate but with even greater rapidity. The cosmos is limitless, and each extension of knowledge opens up new vistas apparently without end. Certainly, we are better trained, better equipped, and better organized for the task than ever before. The only serious obstacles (though they are serious enough!) seem to be man's own limitations—the possible incapacity of his intelligence to understand and to use aright the mighty powers he has invoked.

How mighty are these powers, how incalculable these possibilities, can be grasped only if we realize the full implications of modern science and strive to raise our vision above the paltry concept that limits the scientific movement to material applications of knowledge like mechanical inventions and aids to our comfort or prosperity. As one scientific writer well observes, "It is this aspect of science that has captured the popular imagination. But science has gone deeper.

5. Professor Horace Lamb: Presidential Address before the British Association for the Advancement of Science, August 1925.
6. W. C. Curtis, *Science and Human Affairs,* p. 3 (New York, 1922).

Human thinking has been revolutionized by scientific knowledge and method. It is this spiritual aspect of modern science that is its most significant feature. By comparison, the material aspect is insignificant.... Modern thought is the outcome of modern science. The scientific habit of mind is the unpredictable factor in the life of mankind today. Its possibilities for the future defy all estimates of prophecy."[7]

The combined effect of these material and spiritual developments upon every phase of our daily lives is simply beyond calculation. If we really thought about it, we should be filled with constant fascination and amazement. The fact of the matter is that we are living in the midst of a revolution more intense and far-reaching than all the mere political revolutions that the world has ever seen. Scarcely a year passes in which our conditions of life, our modes of conduct, and our very habits and amusements are not subtly yet surely altered. We live in a world of kaleidoscopic change and are constantly forced to adjust ourselves to unprecedented circumstances. Indeed, it is this very necessity for continuous readjustment, this perpetual impact of novel stimuli, that is in great part responsible for the restless, hectic, discontented temper of our time. Dimly, even the least reflective of us sense the remorseless action of mighty and disturbing forces, of which we are at once the authors and the subjects.

From this transforming action of modern science nothing seems exempt: not even our basic concepts of the universe. Unless we stop to ponder it, we can hardly grasp the amazing change wrought during the past generation in our ideas regarding such fundamental concerns as Matter, Space, and Time. The nineteenth century had apparently charted the cosmos in fairly definite fashion. Our world seemed a substantial affair made up of indestructible atoms, run by indestructible energy, and governed by "iron laws." Such a world tended to produce an attitude of mind varying from quiet assurance to somber certitude. The prevailing trend of philosophic thought was towards a dogmatic materialism. Above all, novelty and mystery seemed almost banished from the world. Nineteenth-century science having sketched the broad outlines of the cosmos, the task of future generations seemed destined to be mainly a filling in of details.

How the discoveries of the twentieth century have altered all this! "Matter," in the traditional sense of the word, has disappeared, and our seemingly solid world (including our own bodies) resolves itself into countless systems of protons and electrons, pithily described as disembodied charges of electricity, constructed somewhat on the pattern of our solar system and moving with prodigious rapidity. As for Space and Time: instead of being independent and absolute entities, they are coming to be regarded as merely aspects of something yet more fundamental, which is neither Space nor Time, "but a kind of union of them, called Space-Time, having four dimensions, of which Time is the fourth. And then, finally, to add to the marvel, we are told that perhaps not only are Space and Time very much interwoven with each other, but Space, Time, and Matter are all parts or aspects of the same primeval reality, Matter being, so to speak, spots or irregularities or

7. Curtis, *op. cit.*, p. 4.

hummocks or kinks in the Space-Time continuum."[8]

Here, indeed, is a world differing vastly, not merely from traditional beliefs but even from the scientific concepts of a generation ago. We stand bewildered at the shattering of what we had considered the most firmly established assumptions concerning the material universe.

Yet this bewilderment is mingled with a subtle sense of exultation. These sudden extensions of human knowledge disclose novel and illimitable vistas! A new world, nay, a new cosmos, bursts upon our gaze, and those in the van of scientific thought experience the challenging thrill of mighty discoveries and glorious adventures. The set and somber outlook of nineteenth-century materialism wanes, and the banished spirits of novelty and mystery return.

The effect upon our general attitude toward life is as striking as it is significant. Compare the zest and almost romantic freshness displayed by our scientific pioneers with the pessimistic gloom shown by so many leaders in politics, literature, and other traditional fields of human activity. So must Columbus and his fellows of the great age of geographical discovery have stood out against the background of their time, bringing to the cramped and jaded Medieval Age the tonic tidings of new worlds and marvels hitherto untold!

The growing popular appeal of science is well described in the following lines: "There is a fascination in watching a moving thing, especially if it is moving fast. Science is moving fast nowadays, making more progress than ever before, gaining more ground in the last quarter-century than in any previous century. And science, unlike most human activities, is always moving on to something new, ever advancing into the unknown. The dramas of human life, as they are told to us in novels and newspapers, are based on a few simple plots, endlessly repeated from generation to generation, with slight variations. One gets to know them by heart and gets tired of them before he grows very old. Fashions in dress, arts, and customs recur in cycles. But in the sciences and in the inventions and industries based on them, there is genuine movement and perpetual progress. The chemist makes new compounds, and the biologist develops new forms of plant and animal life such as never existed before. The astronomer discovers unknown stars. The physicist formulates new laws and strange conceptions. The scientist is always turning over a new leaf. The Book of Nature is issued only in uncut editions, and the scientist has to open its pages one by one as he reads."[9]

This adventurous, pioneering spirit of present-day science and its increasing impact upon ever-widening circles of public opinion are among the most heartening aspects of our time. Of course, the "optimism" of the scientist must not be confused with the ignorant and fatuous complacency that so often passes under that name. No one realizes better than the scientist the prodigious alternatives of good and ill that confront our age: the complex problems of adjustment to new surroundings; the equally difficult question of harmoniously reconciling our new knowledge to our traditional attitudes and beliefs; the ominous possibility that

8. G. T. W. Patrick, *The World and Its Meaning,* p. 69 (Boston, 1925).

9. E. E. Slosson, *Keeping Up with Science,* Introduction, pp. v-vi (New York, 1924).

human nature may be inherently unequal to the task. What sustains the scientist's hope is the thought that in a world so plastic and dynamic as he now sees it to be, the mind that has conceived and evoked these problems possesses the wit to solve them; that, despite all the perils that may encompass him, man has a good fighting chance of winning through.

That is the challenging and inspiring note of contemporary science, compared with the more somber tone of a generation ago. As one writer racily puts it, "Life, in short, has become a solemn sporting proposition. Solemn enough in its heavy responsibilities and the magnitude of the stakes to satisfy our deepest religious longings; sporty enough to tickle the fancy of a baseball fan or an explorer in darkest Borneo."[10]

Once again, we confront the extraordinary paradox of modern man and his civilization. Consider this civilization, with its foundations laboriously laid in past ages and its almost jerry-built superstructure shot up in recent times. Can its architectural strains and stresses be composed into a stable and harmonious edifice? Consider likewise ourselves, peering into the depths of the cosmos, yet bearing within us the limitations, the passions, and the irrationalities of savage and even of pre-human ancestors. Can such eminently "unscientific" beings really adjust themselves to their new scientific world?

---

10. J. H. Robinson, *The Mind in the Making,* p. 213 (New York 1921).

OUR UNSCIENTIFIC SELVES

Chapter II

Our Unscientific Selves

Perhaps the most impressive development of modern science is the new knowledge that we have acquired concerning human nature. It is no overstatement to say that scientific research has revolutionized our ideas concerning ourselves even more profoundly than it has our outlook on the starry heavens above us or the earth beneath our feet. For the first time in his history, man is beginning to *know himself*—the indispensable starting point for the wise handling of human problems.

Let us consider the amazing difference between the traditional and the scientific concepts of human nature. In the past, man has been usually regarded as a rational, sensible being, tending normally to prefer a prosperous, orderly, and peaceful life if he were placed in fitting surroundings and if the advantages of such an existence were properly impressed upon him.

Philosophers acclaimed the power of "reason," economists depicted their common-sense "economic man," while statesmen urged the transcendent value of liberal institutions. Even today most of our political and social efforts are quickened by the belief that if mankind can only get enough wealth, leisure, security, justice, and similar favorable conditions, all will be well and peaceful progress will be automatically assured.

Now, however, along comes the scientist with some decidedly disturbing observations. Viewed objectively, man appears quite different from what we had fondly imagined him to be. No doubt, man possesses all the hopeful attributes above described, but he also possesses many others of a very different character. In short, the picture hitherto painted has been altogether too simplified and one-sided. The truth of the matter is that man is a walking paradox, a baffling compound of diverse tendencies. He is the product, not merely of a few thousand years of civilized and semi-civilized surroundings, but of a race-life extending back through at least half a million years of barbarism and savagery to untold eons of pre-human existence. And, as a result, all sorts of primitive characteristics persist alongside his higher and more recently evolved mental and moral qualities.

Here we glimpse the endless paradox of human nature. Man, to be sure, possesses a marvelous brain and an even more marvelous moral sense. He is capable of the profoundest reflections and the loftiest ideals. But he cannot do such things altogether or all the time. The poetic simile of two natures struggling within oneself is an artistic perception of what is forever going on in every human life. Thus, man is at once the sensible, rational, moral, and farsighted being that we have imagined him—and at the same time an irrational, passionate, restless creature "who naturally loves to hunt, wander, brawl, intrigue, drink, play, dance, and sing, take risks, and above all preserve the sense of his own importance."[1]

It is this strange combination of the saint, the sage, the child, the barbarian, the savage, and even the pre-human with which we have to deal in all life's problems. A well-known psychologist, in a book aptly entitled *The Caveman within Us,* puts the matter very well. "Do you know," he writes, "that this Cave creature within you is *at bottom* absolutely unethical, antisocial, egotistical, primitive, and otherwise destitute of all the cherished virtues? In this, there is no cause for alarm. It merely proves that you are human, neither more nor less. And by getting an insight into the situation, you will be better able to adjust yourself to the problems of life. It will become a valuable guide in pursuing your destiny, as necessary in present-day life as is the chart of the navigator on the high seas."[2]

Our paradoxical make-up may not cause us pessimistic alarm, but it certainly should be seriously borne in mind in both our private and public affairs. It is to his brain that man owes his extraordinary rise in nature, and it is our mental and moral qualities that are the hope of the future. We are extremely proud of our superior intelligence, yet it is extraordinary how little we use it and how seldom we try to learn how to employ it more advantageously. When we analyze our actions, we see that they are determined, not so much by rational thought as by emotional processes—instinctive likes and dislikes, prejudices, passions, and other states of *feeling*. What passes for thought is mainly emotional efforts to convince ourselves and our fellows that our actions are wise and reasonable, whereas the truth is that most of them are unreasonable and happen chiefly because we are "built that way."

In fact, our mental life is a deep, swift-flowing, turbid stream, of which we see only the surface. Inextricably bound up with our bodily processes, mental activity goes on largely beneath the level of consciousness; yet that part which is conscious is subtly influenced by what goes on in the depths below. Again, of what does even the conscious part consist? Mainly of spontaneous, undirected brain activity, hardly to be called "thinking" in the ordinary sense of the word, and largely of a highly personal, intimate, trivial, or even ignoble character. This brain activity, known to psychologists as "revery," which goes on ceaselessly, asleep or awake, is our most congenial form of consciousness. From it we reluctantly rouse ourselves to decide upon our voluntary acts or courses of conduct.

Such is the unflattering background of our mental life. Obviously, it differs vastly from the traditional concept. "Thinking," as applied to deliberate, rational brain activity, instead of standing apart in detached and lofty isolation, is seen to be always complicated by irrelevant and distracting factors and is rarely maintainable even on that dubious level for long periods of time.

1. A. E. Wiggam, *The New Decalogue of Science,* p. 159 (Indianapolis, 1922).
2. W. J. Fielding, *The Caveman within Us,* p. 2 (New York, 1922).

Indeed, it is hard to tell offhand how much even of our most apparently serious "thinking" is rational thought and how much consists of subtle emotional reactions. This is especially evident when we are confronted by ideas or arguments that challenge our accepted beliefs and opinions. In such cases there occurs a mental process termed "rationalizing," by which we instinctively attempt to discredit the challenge and justify our attitude. On this point Robinson well says, "We are incredibly heedless in the formation of our beliefs but find ourselves filled with an illicit passion for them when anyone proposes to rob us of their companionship. It is obviously not the ideas themselves that are dear to us, but our self-esteem, which is threatened. ... We like to continue to believe what we have been accustomed to accept as true, and the resentment aroused when doubt is cast upon any of our assumptions leads us to seek every manner of excuse for clinging to them. The result is that most of our so-called reasoning consists in finding arguments for going on believing as we already do."[3]

Furthermore, these mental shortcomings are common to all of us, existing in the very loftiest minds. Being human, even the leaders of the race have in their make-up the same complex, contradictory, irrational traits as their less distinguished fellows, the chief difference being that most exceptionally endowed individuals possess an intellectual grasp and emotional control capable of thinking more clearly and acting more wisely than the mass of mankind.

Nothing better illustrates the rarity of sound, unprejudiced thinking than the difficulty experienced in attaining it even by those most convinced of its necessity. If there be one type of man with whom objective, impartial thought is at once a general ideal and a specific method, it is the scientist. Yet the widespread notion that the man of science is *consistently* scientific on all occasions, both in his professional and private life, is an amusing error. As a well-known British savant ruefully admits, "The ideal of truthfulness is probably more rigorous with the scientist, *qua* scientist, than with any other kind of man. But it would appear that this dispassionate rationality is hardly won and precariously maintained. Outside his laboratory the scientist may, and usually does, show himself as simple, as kindly, as credulous, and as irrational as any other man."[4]

This brief analysis of man's mental and emotional make-up gives us some conception of the difficulties involved in adjustment to novel conditions and new ideas. It shows us why not merely scientific discoveries but also the scientific attitude, with its freedom from bias, its impersonal search for facts, and its resolute quest for objective truth, instinctively arouse widespread hostility. History proves all too clearly that this is the usual human reaction to anything that offends traditional beliefs or prejudices. It is one of the perversities of human nature that despite the experience of ages, which should have taught man the supreme value of creative thinking, he still tends to distrust and dislike pioneering intelligence as something wickedly shocking and disturbing.

This instinctive dislike of disturbing novelty, so widespread in human nature,

3. J. H. Robinson, *The Mind in the Making,* pp. 40-41 (New York, 1921).
4. J. W. N. Sullivan, *Aspects of Science,* p. 101 (London, 1924).

should be seriously pondered, because it is one of the chief obstacles to progress. We are "up against" the toughest problem of readjustment that mankind has ever faced. Never before has the impact of new ideas been so insistent; never before has the pressure of new conditions been so intense. Somehow or other we must genuinely assimilate the vast extensions of recently acquired knowledge to our existing (and mainly traditional) idealistic and cultural scheme; somehow or other we must adapt ourselves to our new scientific world—or suffer disaster. Let us probe a little deeper into our unscientific selves in order to estimate more accurately the emotional resistance that may have to be taken into account.

The instinctive reaction of the individual to disturbing novelty is a mixture of fear and anger, tending to result in an attitude of aggressive hostility. Furthermore, this attitude is apt to spread by crowd contagion and may either flare into outbursts of mob violence or harden into dogmatic intolerance, producing systematic opposition and persecution. The volcanic possibilities latent among even the most civilized peoples should not be forgotten. As one competent student of human nature well puts it, "There is little doubt that, in the main, humanity is still Yahoo-manity. Alike in mentality and in *morals*, modern man is still substantially identical with his Paleolithic ancestors. He is still the irrational, impulsive, emotional, foolish, destructive, cruel, credulous creature he always was. Normally the Yahoo in him is kept under control by the constant pressure of a variety of social institutions, but let anything upset an established social order, and the Yahoo comes to the front at once.... Thus, civilization is not even skin deep; it does not go deeper than the clothes.[5]

There is thus nothing strange or unprecedented in the bitter hostility toward both the scientific attitude and many of its fruits, which we observe not only among the masses but also in more distinguished quarters. On the contrary, we should remember that ours is the first epoch in human history in which objective thinking and scientific research have been even tolerated. In previous epochs the few superior minds who attempted to investigate nature's mysteries were either contemptuously ignored or savagely persecuted. It is only during the last century or two that real freedom of thought and inquiry has been allowed. Throughout the Middle Ages no true scientific activity was possible. The Church was solidly backed by public opinion in its ban upon such pursuits as wicked and diabolical. The fate of rare scientific intellects like Roger Bacon[6] was a warning too grim to be disregarded by inquiring minds. Bacon's pathetic little laboratory was sacked by a fanatical crowd, terrified by the possible consequences of his "black magic," and Bacon's arrest and imprisonment probably saved him from being torn to pieces by the furious mob. This bigoted attitude waned slowly and reluctantly as the Middle Ages gave place to modern times. As late as the year 1633, for instance (less than three centuries ago), Galileo, the great scientist, was condemned by the Inquisition and, on peril of his life, was forced to recant his damnable error that the earth moved around the sun!

5. F. C. S. Schiller, *Tantalus or the Future of Man,* pp. 32-33 (New York, 1924).
6. Roger Bacon, one of the outstanding figures of the Middle Ages, lived during the Thirteenth Century.

Beside Medievalism's complete taboo of science, contemporary protests seem mild indeed. Rarely are they thoroughgoing condemnations, but rather censures of certain aspects of the scientific movement. In truth, hostility to science is today largely an instinctive spirit of uneasiness and dissatisfaction, often rather inarticulate and ill-understood. That this anti-scientific bias is a serious obstacle to the harmonious development of our civilization will be made clear in subsequent chapters, where the specific problems of modern life are considered in detail. Let us here content ourselves with a brief discussion of hostility to the scientific movement as expressed by some of its eminent opponents.

Ever since the rapid development of modern science in the first half of the Nineteenth Century, a chorus of critical disapproval has arisen, perhaps voiced most vehemently by theologians, but also uttered by leaders in literature, art, education, and other fields. The driving force of this opposition is the belief that, despite its material benefits (or, perhaps, just because of them), science is really the enemy of mankind, destroying piety, simplicity, contentment, and beauty, and heading toward either sophisticated barbarism or utter ruin. During the Nineteenth Century, personalities as varied in outlook as Carlyle, Cardinal Newman, Ruskin, and Tolstoi, however much they might differ in their positive programs for human betterment, were agreed in their condemnation of modern science and its fruits. Arnold of Rugby, the celebrated English educator, piquantly voiced this anti-scientific feeling when he wrote to a friend, "Rather than have physical science the principal thing in my son's mind, I would gladly have him think that the sun went round the earth and that the stars were so many spangles set in the bright blue firmament." As able a thinker as the French writer Joseph Reinach contended zealously that scientific research must be stopped in the clear interest of humanity, and contemporary prophets of woe range all the way from Bertrand Russell to Mr. Chesterton and the late Mr. Bryan.[7]

Of course, criticisms of our scientific age are extremely varied, being inspired by widely different motives and directed at very different things. Some protests are plainly emotional reactions due to dislike and fear of novel ideas that shock established beliefs and prejudices. Others are examples of "rationalizing," in which the offended emotions are concealed under an appearance of dispassionate thought. Furthermore, both these sorts of argument are usually complicated by ignorance of what is condemned.

Another type of criticism, however, deserves much higher intellectual consideration. These are the criticisms that clearly spring from objective thinking, uninfluenced (or influenced only slightly) by emotional processes. A good example of this type of criticism is the pessimistic judgment of Bertrand Russell on our civilization.[8] Himself a scientist, Russell bases his disquieting conclusions not merely

7. For the extreme bitterness to which anti-scientific hostility can go, the reader is referred to an English work entitled *The Idolatry of Science,* written by S. Coleridge and published in London in 1920. Herein science and scientists are belabored with extraordinary vehemence.

8. Mr. Russell's thesis is most compactly stated in his book, *Icarus or the Future of Science,* already referred to in the previous chapter.

on certain tendencies in the scientific movement, but also on innate limitations and shortcomings of human nature, which may render modern science ultimately more of a curse than a blessing. This is a thoroughly objective and rational attitude. We have already noted the hazardous elements in our civilization, which must inevitably raise the query whether man is not today undertaking a task beyond his powers and is thus foredoomed to failure.

What the outcome of the great experiment of modern life will be, no one can surely say. However, it is worthy of note that pessimistic judgments like Bertrand Russell's are no longer so frequent in scientific circles as they were a generation ago, the prevailing trend of scientific thought today being basically hopeful in tone. Fairly typical of this attitude are the following reflections of a British scientist: "We cannot escape the structure of our own minds, it is true, but we do not yet know what that structure is; we do not know what barriers are unbreakable; we do not know what thoughts are thinkable by man. A universe in whose construction so plastic and mysterious an entity as the human mind collaborates may very well hold great surprises."[9]

One thing is certain: we are what we are, and we cannot turn ourselves into anything else. Whatever conscious bettering of the human stock may be accomplished in the course of future generations, we living mortals must recognize that we are an illogical blend of high intellectual and moral qualities combined with primitive passions and perversities, which will alternately inspire and vex us in our great task.

Yet, because we cannot shed our shortcomings is no reason why we should be unduly alarmed by them. Our instincts, insofar as they seem to us bad, can largely be controlled or redirected into relatively harmless channels. Furthermore, our social order can be modified so as to be in harmony with human nature instead of thwarting or contorting it. Here, again, we see the vital importance of our new knowledge concerning man, who, as already remarked, is for the first time in his history really beginning to know himself.

How vital this really is to human progress may be better appreciated when we consider some of the ominous aspects and tendencies of modern life, as we shall do in the next chapter.

9. Sullivan, *op. cit.*, p. 144.

THE PERILOUS PRESENT

CHAPTER III

THE PERILOUS PRESENT

DURING THE DARK HOURS OF the American Revolution, Thomas Paine wrote, "These are the times that try men's souls." How aptly those words fit our own days! We live in times far more revolutionary than the mere political struggles of the Eighteenth Century. Our "revolution" affects every phase of human activity, girdles the globe, and involves all mankind. In preceding chapters, we analyzed the paradoxical character of our age: the striking contrast between man and his new scientific world. Let us now briefly survey some of the disquieting trends that flow from this situation.

The outstanding feature of contemporary world affairs seems to be a profound and general disharmony between Old and New, resulting in endless maladjustments, perplexities, and conflicts. The changes wrought in our everyday lives by the disturbing impact of new inventions and methods are merely the outward sign of what is going on in the mental and spiritual realms of ideas and ideals. Keen-sighted observers are coming more and more to realize that our whole civilization is undergoing an acid test of ruthless revaluation. Throughout the civilized world the most cherished beliefs are being scrutinized, while no institution, however venerable, escapes the fire of criticism. So grave is the problem of adjustment, so acute the crisis of transition, that we have good reason to fear lest the social fabric give way and our civilization go down in ruin. We have already seen that its best chance of survival lies in the one factor that distinguishes our age from past times—the spread of exact knowledge, inspired by and infused with the scientific spirit. Further on, we shall discover the full significance of science in modern life. In the present chapter, however, let us content ourselves with a survey of current tendencies, in order that we may better realize the gravity of the problems we have to face.

The profound perplexity of our situation is well summarized in the following lines, from the pen of a shrewd observer of world affairs: "Never before have we had to rely so completely on ourselves. No guardian to think for us, no precedent to follow without question, no lawmaker above, only ordinary men set to deal with heartbreaking perplexity. All weakness comes to the surface. We are homeless in a jungle of machines and untamed powers that haunt and lure the imagination. Of course, our culture is confused, our thinking spasmodic, and our emotion out of kilter. No mariner ever entered upon a more uncharted sea than does the average human being born in the Twentieth Century. Our ancestors thought they knew their way from birth through all eternity; we are puzzled about day after tomorrow."[1]

The dangers as well as the benefits of modern science and "progress" are so patent that they have inspired an imposing literature of warning. For example, there exists a considerable school of criticism directed against the machine aspects of our civilization, ranging from the protests of Ruskin to the Robots of the satirical drama *"R.U.R."*[2] Other critics, while not condemning material progress in itself, assert that its speed has become too rapid for man to adjust himself, and claim that his capacity for adaptation to new conditions is being strained to the breaking point. "The rate of change is in many respects more important than the extent of change... Consequently, it is above all necessary to ask, in the case of man as of the other animals, whether the environment is outdistancing the inherent capacity."[3]

Still other critics, though admitting man's innate ability to solve his present problems, doubt whether he will actually display the intellectual grasp and emotional poise necessary for the task.

We are, therefore, warned against the weakening effects of conflicting loyalties,[4] and the dangers of misguided or exaggerated idealism.[5] The need of scrutinizing the possible effects of even our loftiest ideals is ably stated by Professor Patrick, who writes:

"The special characteristic of our time is, therefore, not the presence of evils, of which, to be sure, there are quite enough, but the peculiar consciousness of them and the resolute will to cure them—a will so persistent and so determined that it is certain that the twentieth century will see profound changes in our social order. But it does not necessarily follow that these changes will be beneficial. They will be experimental, and the prophet of social catastrophe may well question whether the radical experiments that it is proposed to make in social reform may not prove to be destructive of the civilization that we have. This is the first time in history that man has consciously and with determined purpose entered upon the task of directing his own fortunes. Hitherto he has been a puppet in the hands of cosmic forces. ... Now the period of conscious control has come.

"But is this conscious control to be intelligent control? So far as we can see at present, the era of intelligent control lies far in the future, and the control that is to mark the twentieth century will spring from an impulsive idealism characterized by a keen sensitiveness to our present social evils, rather than by a comprehensive grasp of the whole situation. We are to enter upon the deliberate attempt at social reconstruction, but with a kind of adolescent impetuousness and a fatuous, almost

1. Walter Lippmann, *Drift and Mastery,* p. 196 (New York, 1914)
2. Most of this literature is distinctly anti-scientific in tone. For a criticism of the mechanistic aspects of our civilization written in the scientific attitude and with due appreciation of scientific knowledge, see R. A. Freeman, *Social Decay and Regeneration* (London, 1921).
3. Coblentz, *op. cit.,* p. 138.
4. See: C. C. Josey, *Race and National Solidarity* (New York, 1923). Although Josey's book is concerned mainly with the specific problem of national and racial, as opposed to international and supra-racial, ideals, it treats the general problem of conflicting loyalties in understanding fashion.
5. See: William McDougall, *Ethics and Some Modern World Problems* (New York, 1924).

fanatical, faith in the magic of certain symbols to cure social evils. This is, no doubt, a necessary stage in the progress of social control, but it is not without its dangers. We have gained the power to remodel our social order; have we gained the necessary poise and the scientific, historical, and psychological knowledge that will make our meddling safe?"[6]

The importance of emotional, as distinguished from intellectual, factors in the solution of our problems becomes clear when we consider the undoubted increase of emotionalism that has occurred during the past decade. The Great War let loose a cyclone of passion, which has not yet subsided far. Indeed, it may be that the Great War was itself merely a symptom—an episode in a much vaster struggle of ideas and elemental forces, begun long before the war and destined to last long after its close.

However that may be, certain it is that in almost every phase of human activity, from politics to religion, we today witness a disquieting upsurge of irrationality, violence, and intolerance. The dangers of the political trend are thus forecast by H. G. Wells: "It is quite possible," he writes, "that over large parts of the world that we now call civilized, the quality of the collective life may deteriorate from a broad-minded reasonableness towards violence and sentiment. And it may rest at that lower level. The recuperative activities of school and literature and reasonable protest may be sufficiently checked and suppressed to prevent recovery....This increasing violence in public life is not a new thing. It has always been there potentially; it has simply been released through the confusion and enfeeblement of the countervailing forces of education and civilized usage. It is a relapse, a falling in. It is a relapse from which recovery may be a very long and difficult process."[7]

This emotional trend is not confined to the broader aspects of politics, national and international; it is evident also in the relations of the State and the individual citizen. Even among the English-speaking peoples, with their long tradition of personal liberty, we witness a tendency toward greater public compulsion, verging on tyranny. The State, by legislation and police activity, tends to interfere with the private activities of individuals, not only in the general interest of society but also with the idea of constraining the adult individual to certain lines of conduct "for his own good." Such a mixture of what we may call "private" with public motives is evident in the American prohibition movement.

Furthermore, this State interference with what used to be regarded as strictly private concerns is the logical outcome of the growing pressure of public opinion upon private conduct—even upon highly personal matters like dress, taste, and attitudes. In many parts of the United States, especially in the smaller communities, standards of behavior and thought are alike rigidly fixed, and any pronounced departure from these standards is punished by all sorts of petty persecution. The cramping, leveling effect of this intolerant "Main Street" attitude is a serious handicap both to individual development and to true social progress.

6. G. T. W. Patrick, *The Psychology of Social Reconstruction,* pp. 22-23 (Boston, 1920).
7. H. G. Wells, "A Forecast of the World's Affairs," from his article in the collaborative work entitled *These Eventful Years,* Vol. II, p. 11 (London and New York, 1924).

The growing emotionalism of our times involves a corresponding diminution of the rational, tolerant attitude best described as "liberalism." This decline in the qualities of open-mindedness, readiness to hear and discuss other viewpoints, respect for differences of opinion, and downright common sense reveals a truly amazing paradox between the way we accept the material benefits of science and at the same time reject the scientific spirit. Modern science is the natural fruit of an objective, dispassionate mental attitude. Without this attitude, the stream of inventions and processes that have transformed our world could not have arisen, and unless that attitude is maintained and is spread among the general population, the stream will eventually dry up. Let there be no mistake: we may cram ourselves with mechanical marvels like radio sets, motor cars, and dishwashing machines; yet we may be less truly open to the scientific spirit than the lonely pioneer, dwelling in a log cabin and living on parched corn and wild game but thinking his own thoughts and minding his own business.

The emotional, irrational tendencies visible during recent years are nowhere more evident than in the field of religion. In a subsequent chapter we shall discuss the broader aspects of the relations of religion and science. Let us here consider the hostility shown by those who distrust and fear scientific thought because of its effect upon traditional religious beliefs. This anti-scientific hostility (though by no means confined to America) is best exemplified in the so-called "Fundamentalist" movement of the United States. The nationwide campaign of the Fundamentalists to prohibit the teaching of the theory of evolution in tax-supported schools and colleges attracted worldwide attention with the sensational Scopes Trial at Dayton, Tennessee, a State where the Fundamentalist contention had been enacted into law. Fundamentalism's attitude toward science was pithily expressed by its best-known spokesman, the late Mr. Bryan, when he declared that he was "more interested in the Rock of Ages than in the age of rocks." His opinion of scientific speculation is revealed in the following lines, written shortly before his death: "If it were possible to regard the evolutionary hypothesis as merely a comedy, intended to furnish light reading to relieve minds overtaxed by the demands of this hurrying age, I would not care to disturb the sapient scholars, who, seeking out the shady places and cushioned seats, spend their time endeavoring to clothe the bodies of their remote ancestors with hair and trying to remove the hair from the heads of their remote descendants."[8]

To regard Mr. Bryan as an unusual or extreme example of Fundamentalism would, however, be a great mistake. The Fundamentalist viewpoint is held by millions of passionately sincere adherents, while Mr. Bryan was far more moderate in tone than are many others of their spokesmen. So active and successful is the well-organized and well-financed Fundamentalist movement that upholders of scientific thought and academic freedom are seriously alarmed at the possible consequences. Learned societies like the American Association for the Advancement of Science have bestirred themselves, and a regular fighting organization, the Science League of America, has been formed to combat Fundamentalist propaganda. This

8. W. J. Bryan, "Mr. Bryan Speaks to Darwin," *The Forum,* July 1925, p. 106.

League, sponsored by many of America's leading scientists and educators, has made a careful study of Fundamentalism's aims and attitude. How violent Fundamentalists can become may be seen by the following excerpts drawn from one of the League's publications:[9]

The Rev. John Roach Straton, of Calvary Baptist Church, New York City, who is probably the most outstanding Fundamentalist leader since the death of Mr. Bryan, roundly declares, "Better wipe out all schools than undermine belief in the Bible by permitting the teaching of evolution." The Pasadena, California, *Broadcaster*, a Fundamentalist organ, remarks editorially, "When a teacher in the public schools or colleges talks about evolution, take him out and put him with this class that he claims to be the offspring of. Southern California has a number of fine zoos, and we believe, as Dr. Riley,[10] that it would be better to have them take away the eyes from the physical body of the children of today than to blind their spiritual eyes with such rot as is being taught over this land of ours." A Kansas City minister, speaking at a Christian Endeavor meeting in Denver, announces, "Worse than an assassin who kills the body is he who shatters the faith of youth," and goes on to say that rape fiends are burnt, but they are saints in comparison with the teachers of modern science. Another Fundamentalist organ, *The King's Business,* published by the Los Angeles Bible Institute, in denouncing a murderer, asks that he be hanged and then goes on to attack "those highly educated preachers and professors who are a thousand times more dangerous to our country" and demands that "equal justice be meted out to the criminals in school and church."

Commenting on the significance of the above quotations, the League says editorially, "Anyone who has faced, as the president of the Science League has done recently, five thousand furious Fundamentalists, laughing aloud at the simplest scientific statement, and roaring and howling their rage at the slightest opposition to the ignorant and prejudiced statements of their spokesmen, will realize the grave danger involved in such utterances as those just quoted. Every one of these Fundamentalists is a voter, and they will all vote against evolution. We shall be fortunate if they do not carry the lynching spirit incited by their speakers and writers to an even more dreadful point than the passage of anti-evolution laws. ... No cultured Roman of the second century would have believed the Dark Ages probable; yet they came, and civilization went down with them. We are faced with a threatened return to a new Dark Age, and only strong organization and whole-hearted cooperation of the enlightened public will prevent it."

This apprehension is shared by many other scientists and educators. As Chancellor Hadley, of Washington University, St. Louis, stated in a public address, referring to the Tennessee anti-evolution law, "I believe that with a vigorous effort such a law or proscription can be, and will be, established in fifteen or twenty states. The next step will be to banish such teaching from all schools, and we may witness the spectacle in this country of men and women facing prosecution who decline

9. Pamphlet entitled, *A Call to Action!* published by The Science League of America (San Francisco, 1925).

10. The Rev. W. B. Riley, founder and executive secretary of the World's Christian Fundamentals Association. A leading organizer and spokesman of the Fundamentalist movement.

to accept the literal statements of the Old Testament upon geography, geology, astronomy, and the origin of human life."

Whether or not these judgments are unduly pessimistic, the violent fanaticism so widespread in the Fundamentalist movement certainly emphasizes the point with which we are here immediately concerned; that is to say, the volcanic emotions latent among even the most civilized peoples and the apparent increase of emotionalism in the public's attitude toward current affairs.

These disquieting trends in present-day life emphasize the query that we have already raised: the question of whether we have the innate ability to handle our modern civilization. Do we really possess the combination of intellectual grasp and emotional poise needed for the task? That is certainly a moot point. As that wise observer, the late President Stanley Hall, remarked, "Man has not yet demonstrated that he can remain permanently civilized."

So disturbing is the impact of new forces and ideas that they may destroy the old social controls without creating new ones and may thus disrupt our social fabric through sheer overstrain, irrespective of the theoretical excellence of the proposed changes. "This is not to say that any of the old ideas or the old institutions are perfect or holy, or even good. It is only that the obedience to laws and the restraint and self-control that are necessary for social order have been associated with these things in the human brain. A wholly new set of motives for social order is perhaps conceivable, resting upon none of these old institutions; but the human brain changes slowly, and an entire civilization might collapse in the process of a crude and reckless attempt at readjustment."[11]

That this danger is a very real one becomes evident when we examine current social and economic trends. Here we see the same tendency toward violence and irrationality that we have already observed in politics, religion, and other fields. We are not here discussing the respective merits or demerits of the present social order and of the various reforms proposed. What we are concerned with in this chapter is the mental and emotional attitude displayed. And surely, we cannot fail to observe a strong tendency to extreme emotionalism, alike among the critics and the defenders of existing economic and social institutions. The reformers seem so fired by the sense of present evils that they tend to embrace violent gospels like Communism or "Bolshevism," disregarding, or even denouncing, the plain teachings of modern science on such basic matters as the physical world and human nature. On the other hand, conservatives seem to be so filled with fear and hatred of these radical proposals that they are apt to be swept into blind reaction, in which the need for any real change is denied and the new scientific knowledge is equally ignored. Both these attitudes are, of course, essentially emotional, devoid of constructive thinking, and serious obstacles to genuine progress. Here, again, we witness eruptions of primitive passion, which may overwhelm reason and kindle destructive social conflagrations.

Our observations on the power of the emotions throw into bold relief the vital importance of the mental factor in human affairs. It is rational intelligence,

11. Patrick, *The Psychology of Social Reconstruction,* p. 204.

displayed in clear, cool, constructive thinking, that is our safeguard against the destructive forces of passion and unreason. Only by genuine thought can our problems be faced, understood, and solved.

Unfortunately, this is not as easy as it sounds. We have already seen the difficulty of objective, dispassionate thinking even among persons of high intelligence. Furthermore, one of the disquieting discoveries made by scientific studies of human nature is the fact that such persons are not numerous. We may as well recognize the unpleasant truth that high intelligence is rare, most persons being of only mediocre mental caliber, while a distressingly large number are distinctly low-grade. The mental differences between men, ranging as they do all the way from the genius to the idiot, are vast. Neither physically nor emotionally do we find any such range of variation.

The reason for this is clear. Our physical development goes back many millions of years to the very origin of life on earth. Our emotional evolution is almost equally ancient, most of our emotional processes being plainly discernible among the higher animals. But the special traits of intelligence that distinguish man from all other forms of life appeared only a few hundred thousand years ago, evolved slowly, and have not yet attained the stage of widespread high development. Biologically speaking, therefore, high intelligence is a very recent human trait, which is still comparatively rare.

Now intelligence, like all other traits, is inherited. But, being a relatively recent quality, as yet developed only rarely to a high degree, it is biologically unstable and subject to easy retrogression. Considering its comparative rarity, high intelligence might quite conceivably be bred out of the human race. That would be an appalling calamity, because it is to the small minority of highly intelligent individuals that mankind owes its advance athwart the ages. Every new idea, every mechanical invention, is the product of superior minds, and the vast accumulation of knowledge and power, slowly amassed in the past and culminating in the splendid present, is in the last analysis due to humanity's élite—that infinitely precious minority in whose hands lies the destiny of the race.

Surely, never was high intelligence so needed as in our complex modern civilization. Yet the weight of scientific evidence apparently reveals the alarming fact that this most vital quality is today on the decline. All over the civilized world the conditions of modern life seem to handicap the most intelligent elements, whose numbers are stationary or diminishing, whereas the low-grade and mentally defective elements of the population are rapidly increasing. The problem of human degeneration is literally a life-and-death question, involving not merely the fate of modern civilization but possibly the entire future of mankind.[12] Certainly, the result of continued racial impoverishment would be the decline or collapse of our civilization and the coming of a new "Dark Age."

This melancholy anticlimax to the brilliant drama of modern life is a specter that today haunts many observers of our time. Indeed, some go so far as to predict

12. I have discussed the problems of degeneration, race impoverishment, and their consequences at length in my book, *The Revolt against Civilization* (New York and London, 1923; Dutch translation, 1924; German translation, 1925; French and Italian translations pending).

that the downfall of our civilization would be but the prelude to the extinction of the entire human race. The noted British scientist, Sir E. Ray Lankester, in his essay entitled "Nature's Insurgent Son,"[13] describes Man as the arch-rebel against Nature, who has so defied natural processes in the creation of his artificial environment that a return to the "state of nature" is impossible since it would spell nothing short of racial death.[14]

If this judgment be correct, humanity's watchwords must be "No Surrender!" and "Carry On!" Only by fresh defiance of nature and only in greater mastery over natural forces and a more perfect artificial environment can man find safety and assure his future. And he can do this only by using to the full the brains he has inherited and the knowledge he has acquired.

13. *In The Kingdom of Man* (London, 1920).
14. See also: Coblentz and Freeman, *op. cit.*

SCIENCE AND EVERYDAY LIFE

Chapter IV

Science and Everyday Life

If Rip Van Winkle could have slept a hundred years and then awakened in a modern city apartment, he would certainly believe himself in another world. His first sensation would probably be a delightful perception of the extreme comfort of the bed on which he lay, with its box spring and other up-to-date adjuncts to bodily ease. The morning perchance being frosty, he might be reluctantly making up his mind to quit his cozy nest when a being in odd garments would enter, close the windows, and turn a lever beside the wall; whereupon, with no sign of kindled fire, a gracious warmth would soon suffuse the room.

Arising, Rip would perceive an adjoining bathroom, immaculate in white glaze, where hot and cold water gushed at will, and where he would discover all sorts of novel toilet conveniences. Proceeding to the dining room, he would find upon the table strange little mechanisms, attached to the floor or ceiling by cords, that boiled his coffee and browned his toast, while if he peeked into the kitchen to find out how the rest of his breakfast was being prepared, he would see an equally fireless stove, the cooking being done by a mysterious force termed by his hosts "electricity."

Leaving the apartment, he would be shot downward in a moving cage to the street level, where he would enter a marvelous horseless carriage, softly purring and vibrating with life, which would carry him swiftly and silently over shiny black roads crowded with similar machines through endless streets lined with incredibly tall buildings. Everywhere the noise of mechanism and the hum of power would sound in his ears.

Perhaps his hosts would take him to the waterfront, where he would gaze upon huge craft, without sails, yet moving majestically against wind and tide. From the docks he might be borne to a vast building filled with snorting steel monsters dragging files of luxurious coaches, one of which would whirl him along a railed pathway out into a countryside only slightly less mechanized than the city that he had left behind.

Quitting his fire-coach, he might be motored to a wide field whereon stood the strangest sight of all—a metal bird, with rigid wings and roaring song, which would bear him aloft with incredible speed, depositing him near a luxurious country mansion filled with new marvels, where he might spend the evening listening to the voices of long-dead singers or hearing strains of music borne to him through the air from an orchestra performing a thousand miles away.

When Rip Van Winkle lay in his comfortable bed that night reviewing the startling events of the day, what were his reflections? Endowed with considerable native shrewdness, Rip would realize that he had passed through three distinct states of mind. His first reaction was one of amazement, tinged with awe. His hosts appeared to him virtually as demigods, endowed with supernatural wisdom as well as equipped with superhuman power.

Before long, however, doubts of their divinity crept into his mind. He soon discovered that they really knew very little about the marvels they so freely used. They could show him how to press a button or turn a lever and could predict what would follow. But to his questions of *why* this or that happened and what lay behind the immediate process, they could rarely return satisfactory answers.

The general conduct and attitude of his hosts contributed even more to his disillusionment. Neither their thoughts nor their feelings seemed on a perceptibly higher plane than his own. Listening to their private conversation, he found it about as personal, trivial, and short-sighted as that of his contemporaries of a hundred years before. By lunchtime Rip had decided that his hosts were just a lot of ordinary folk enjoying, by some mysterious stroke of luck, the fruits of other extraordinary minds, and he spent his time between soup and coffee wondering how they ever managed their complicated world without getting continually snarled up.

Before evening, however, Rip had again changed his mind. His hosts, who had seemed so like himself, proved on closer inspection to be quite different. Perhaps he could not quite lay his finger on the difference, but he sensed it unmistakably just the same. Finally, he concluded that it was not a conscious but an unconscious attitude; not a direct but an indirect effect of their extraordinary new world—subtly yet surely affecting every phase of daily life, constantly altering everything from habits to beliefs, and exerting a continuous, even though unappreciated, pressure of readaptation. Rip fell asleep uneasily, wondering how it would all end—and very likely wishing himself back in the simpler, quieter life of a century before.

Does not this imaginary episode illustrate the paradoxical position of science in our everyday life? Those engaged in scientific pursuits may have a clear idea of the nature and scope of modern science, but the average man ("the layman," as he is often termed in scientific parlance) has barely the faintest conception of what it is all about. Jostled though he is by new inventions, the layman remains indifferent to the deeper aspects of the changes that are going on. When the average man "tunes in" his radio, boards an ocean liner, or gets a "shot" of anti-typhoid serum, does he think in terms of ether waves, thermodynamics, or bacteriology? He does not. He simply accepts these marvels, not only knowing next to nothing about them but also not even showing much curiosity concerning them. In other words, he just takes them for granted and devotes his attention to his petty personal affairs, precisely as his ancestors did before him.

Nevertheless, he is not the same sort of person as his ancestors. For even though our average man does not consciously think or care about them, these scientific discoveries and inventions are *doing things to him.* They are forcing him to adjust himself to novel conditions; they (though he may not know it) are altering his very philosophy of life. Even the broadest generalizations of science may suffice to change one's idealistic and cultural attitude. For example, out of the vast accumulation of technical knowledge concerning the heavenly bodies, a man may glean only the general idea of our earth's material insignificance in the universe. Yet that single concept, once firmly grasped, can work surprising changes in his mental perspective, causing him to see both the world and himself very differently than he did before.

And of course, where a scientific discovery or invention is at all popularly appreciated, the practical effects alike upon thought and conduct can be truly amazing. Wiggam cites a good instance of this. Speaking of the momentous consequences likely to flow from widespread understanding of the possibilities of race betterment, he writes in a humorous vein: "Some people have little faith in what can be accomplished by this sort of education. But let us consider the amazing thing that *has* happened from educating people about another great discovery of science, the discovery of microbes. Nobody ever saw a microbe with the naked eye, yet its discovery has changed nearly all human life. The education of the people about microbes is one of the most spectacular things in all history. A generation ago microbes were just as mysterious to the public as germ cells and chromosomes are today. Yet every schoolchild now knows about them. They have changed the architecture of our houses, the kind of clothes we wear, and the sort of food we eat. They have changed our schoolbooks, our morals and habits, and even our religion."[1]

The pity of it is that popular appreciation of science is still so rare and so haphazard; that the changes that are taking place in modern life are so unconscious and so ill-understood. As another keen observer of present-day trends well says, "While our information in regard to man and the world is incalculably greater than that available a hundred, even fifty, years ago, we must frankly admit that the knowledge is still so novel, so imperfectly assimilated, so inadequately coordinated, and so feebly and ineffectively presented to the great mass of men that its direct effects upon human impulses and reasoning and outlook are as yet inconsiderable and disappointing."[2]

The need for a truly scientific attitude as a general method for everyday life cannot be overestimated. It is one of the ironies of our time that, while the material applications of science to our comfort and luxury are complacently accepted, the scientific spirit that has made these possible is ignored or even hated. Yet scientific knowledge, infused with and directed by the scientific spirit, is alone capable of composing the disharmonies and solving the problems of modern life. Those people who first grasp and act upon the basic principles of the new knowledge will surely (and rightfully) be the heirs of the future.

In the light of all this, let us briefly survey the factors making for and against the popular appreciation and acceptance of science as a general attitude and

1. A. E. Wiggam, *The Fruit of the Family Tree,* p. 298 (Indianapolis, 1924).
2. J. H. Robinson, *The Mind in the Making,* p. 158 (New York, 1921).

method in everyday affairs.

One of the most hopeful signs of the times is the increasing popular appreciation of the value of scientific research. Only a few years ago a stock joke of the newspaper comic supplements was the bearded, bespectacled "Professor," absent-mindedly blowing himself up in his laboratory or idiotically chasing bugs with a butterfly net. Scientists were popularly regarded as harmless cranks, engaged in quaint but quite profitless hobbies. To be sure, it had to be admitted that, in his abstract studies, the scientist did occasionally hit upon something useful; yet such discoveries were deemed mere chance by-products, too rare to justify the time and labor expended in his abstruse researches. Again, a clear popular distinction was current between "scientist" and "inventor." Of course, the inventor was likewise an odd person, but he was engaged in a specific task with a useful goal, whereas the scientist pursued knowledge for its own sake without thought of any practical application of his intellectual discoveries. In other words, a definite line was drawn between "pure" and "applied" science, the former being popularly disparaged as much as the latter was approved.

Today, popular contempt for "pure" science is distinctly on the wane. This is so mainly because the average man is coming to realize that research can "deliver the goods." The layman is learning that great industries and professional activities, which minister to his most vital needs, spring from scientific discoveries that at first interested only a few learned specialists with no necessary idea of their practical applications. A good example of this is Pasteur's early studies of bacteria, which now give us pure milk, canned foods, refrigerated meat and fruits, protection against disease, and much more besides. Without this single discovery, both the death rate and the cost of living might easily be twice as great as they now are among our congested city populations.

Of course, popular appreciation of scientific research is as yet far from being what it should be. Yet, in one field after another, its importance is being impressed upon the public mind, and these impressions, random and vague though they now are, may in time link up into a consistent, definite attitude.

How necessary such an attitude is for future progress is well explained in the following lines: "We commonly fail to recognize the nature of the scientific studies that have preceded the discovery of many facts applicable to our daily lives. The scientific discoveries of the past constitute the foundations of life in civilized communities as well as of modern industry, and the scientific discoveries of the present will, if we do our share, be as vital in the lives of future generations. We should eliminate from our minds the terms *"pure"* and *"applied"* science. This distinction is a false one, because no line of separation exists. On every hand, discoveries of a theoretical and general nature are of practical value, and, conversely, practical achievements are a constant stimulus to further investigations along theoretical lines. These in turn influence practice in new and wholly unexpected ways....Facts apparently remote from present needs come to be the very lifeblood of subsequent generations. In our quest for facts, we must do so. advance that no spot is left wholly unexplored, for we cannot tell what importance any part of the field may assume. We cannot afford to concern ourselves today merely with what seems useful, since the most important advances of the past have commonly been made through fields that at first gave small promise of value."[3]

Indeed, these very possibilities of novel and marvelous discoveries lend the scientific movement a zest, a charm, a "human interest," quite apart from utilitarian considerations. The adventurous, pioneering spirit evoked by present-day scientific research awakens a responsive thrill in the layman's breast as soon as he becomes aware of what is going on. The sense of far-flung horizons and the feeling of "almost anything may happen" kindle a downright sporting glow that, once felt, rarely cools.

The emotional appeal of science is, in fact, far greater than is popularly supposed. Instead of being a cold, dry, depressingly practical affair, as is usually assumed, science offers cultural and artistic allurements seldom appreciated. As one scientist well puts it, "The study of such sciences as astronomy, physics. or biology awakens emotions not readily distinguishable from those evoked by even the greatest works of art. It is as if the universe with which science deals was itself a work of art; it is, to an increasing number of persons, the greatest of all works of art. Such students often acquire a new standard of aesthetic excellence.... The young William Thomson, reading at night in the library and drawing great breaths of rapture over Lagrange's *Mécanique Analytique,* was experiencing emotions probably not very different from those of Swinburne when reading Shakespeare."[4]

While the aesthetic and cultural aspects of science may as yet appeal to few, it is certain that general interest in scientific matters is today both lively and widespread. This distinctly "popular" interest in science is as unprecedented as it is significant. Nothing like it has ever been seen before. Down to comparatively recent times, science was the affair of cloistered scholars, quite out of touch with the general public. They might now and then be patronized by enlightened sovereigns, as Leonardo da Vinci was by the princes of Renaissance Italy, but the fad remained confined to kings and courtiers. Only by the Eighteenth Century did interest in scientific matters spread to the cultured upper classes, when it became part of a gentleman's education to have at least a bowing acquaintance with the dawning science of the time.

How different is the situation today! Science has emerged from its long obscurity and has become one of the vital topics of our age. The very word *Science* provokes instant attention in almost every walk of life. Editors know that scientific news items make good "copy" and run special scientific articles in the Sunday supplements. Magazines establish "Science Departments" or even devote themselves wholly to a popular presentation of various aspects of scientific progress. Quantities of books are published to serve the same end. Even abstruse scientific studies like Einstein's Theory of Relativity, which could not be really grasped by one layman in 10,000, arouse great popular excitement and provoke endless discussion and animated controversy.

3. W. C. Curtis, *Science and Human Affairs,* pp. 243 and 261 (New York, 1922).
4. J. W. N. Sullivan, *Aspects of Science,* pp. 38-39 (London, 1924).

At first sight this situation might seem all to the good. What could be more heartening than the prodigious extension of popular interest in science during recent years? May it not portend the speedy conversion of the general public to a sound scientific attitude?

Unfortunately, closer scrutiny of the situation shows that no such happy outcome is likely—at least in the immediate future. Favoring factors do exist, as already described, but there are likewise adverse factors, which we will now consider.

These adverse factors inhere both in the public attitude toward science and in the trend of the scientific movement itself.

In the first place, much of the current popular interest in science is of a superficial or controversial character. People may gape at a few scientific marvels or may strike a scientific pose as a fad or fashion. But open mouths soon close, while fashions quickly pass, leaving matters not one whit better than before. To fancy that such postures mean any real appreciation of science or any genuine adoption of the scientific attitude of mind is sheer illusion.

Again, is there not a great deal of current popular interest in scientific matters due to mere controversy, actually inspired by anti-scientific prejudice? For example, the fanatical Fundamentalist is, in a sense, "interested" in science, but his interest is certainly not helpful to scientific progress nor is it conducive to scientific habits of thought. In fine: if we abstract from popular interest in science all superficial and controversial elements, how much that is solid and constructive remains?

This point becomes all the more noteworthy when we consider the tendency, inherent in the scientific movement itself, to become more complex, more detached from our cultural life, and more difficult of comprehension by the layman. That is the inevitable penalty exacted for the prodigious extensions of human knowledge that have occurred during the last few generations. Back in the seventeenth and eighteenth centuries, an intelligent, well-educated man could acquire a fairly good general survey of the entire scientific field. Even in the opening decades of the nineteenth century, an intellectual giant like Alexander von Humboldt might accomplish the task. Today this appears impossible. No human mind can apparently attain an intellectual peak so high that it can thence sweep with its single vision the far-flung horizon of the known. Therefore, instead of the unifying syntheses produced by the great scientific intellects of the past, we must content ourselves henceforth with partial or cooperative surveys, which can hardly be as satisfying or as stimulating.

Furthermore, this extension of scientific knowledge as a whole has been accompanied by a growing complexity of detail in its several branches, rendering it increasingly difficult for the mere layman to understand even one of the many scientific fields. The upshot is that scientific progress tends to become hidden from the public gaze behind a veritable smokescreen of technical terms and involved abstractions.

How this works out in a highly technical branch of science like mathematics is amusingly described as follows: "A mathematical theorem occupying one line of type might very well occupy a volume if written out in ordinary prose in which no terms were used that were not common property. For this reason, modern mathematical discoveries, except in very special instances, cannot be made intelligible except to mathematicians. To learn the language of a highly developed science like mathematics takes about as long as to learn Chinese, but the task of translating into English is very much harder. For this reason, mathematicians cannot hope for intelligent popular recognition; they must be content to be regarded either as vaguely impressive figures or else as mild lunatics busied with incomprehensible and probably trifling abstractions."[5]

This growing complexity of science and the increasing difficulty of its comprehension even by the intelligent, educated layman account for much of the popular confusion about scientific matters. Nothing is more paradoxical than the immense interest in and the immense ignorance of science, which so amazingly coexist in the public mind today. Everywhere we hear people talking blithely and fluently about science, expressing opinions and expounding theories couched in impressive technical terms. It sounds fine—until we discover that most of the speakers do not know what they are talking about but are merely pouring forth a dreadful hash of half-truths, misstatements, and garbled information picked up haphazardly from other persons as ignorant as themselves or swallowed whole from some fakir posing as the high priest of a pseudo-scientific cult.

The fact of the matter is that, in multitudes of minds, the sprouting crop of true scientific ideas is choked by a rank growth of *pseudo*-science, shooting up like weeds in a garden and threatening to kill off the good seed of genuine understanding. To this baneful growth of fake science is due much of the disillusionment, pessimism, and downright hostility to the very name of science that are so unfortunately prevalent today.

Now this profound and widespread confusion in the public mind concerning science and its works must somehow be allayed if we are to have any assurance of true social progress. Somehow or other the intellectual chasm that yawns between the scientist and the layman must be bridged, unless our civilization is to fall into the abyss of ignorance, passion, and unreason. In the last analysis, it matters little how busily professional scientists delve into the unknown and amass rich stores of fresh knowledge if these stores remain unappreciated and unused by ourselves. How vain would be the labors of coal miners, heaping up mountains of "black diamonds," but with no means of distributing the precious fuel throughout the land, to be transmuted into heat, light, and power for our factories and homes! Those carbon mountains might tower to the very heavens, yet we should freeze and starve in cold and darkness, just the same.

How, then, can the public mind be adjusted to the vast extensions of scientific knowledge—growing ever vaster and more difficult of comprehension from year to year? We may as well realize that the average layman (even the well-educated layman) cannot, at first-hand, acquire even a general understanding of modern science, while he is still less able to keep up with the continual stream of new discoveries in the various scientific fields.

5. Sullivan, *op. cit.*, p. 14.

The only way out of the dilemma seems to be through the development of what, in the truest sense, can be called "popular science": not superficial, inaccurate makeshifts to half-satisfy the public demand, but real enlightenment, based on sound scholarship, yet uttered in a clear, intelligible, and stimulating fashion. What we need are men who can follow our scientific pioneers to the remotest frontiers of the Known—and can then come back to tell us in simple, everyday speech the practical meaning of what they have seen.

The urgent need for scientific "popularization," in the true sense of the word, is being increasingly felt and pondered by forward-looking minds. The problem has nowhere been better stated than by one of the keenest students of modern life, President Glenn Frank of Wisconsin University, who calls for a distinct type of popular scholar, termed *The New Encyclopedists.* Developing the idea, he writes, "I suggest that we need to have done for modern knowledge something analogous to what Diderot and the Encyclopedists did in the eighteenth century. To be specific, I think Western civilization would profit vastly from the labors of a group of men who would go, with conscientious care, through the findings of modern biology, psychology, anthropology, experimental ethics, genetics, economics, sociology, chemistry, physics, and all the natural and social sciences, pulling out, tabulating, and reducing to easily understandable terms the net social and spiritual contribution that each of these adventures of the modern mind has made to the future of our civilization. This would give us something approaching an inventory of the raw materials of social renewal upon which we must depend.

"All of these sciences have lying, relatively unused, in their laboratories certain socially usable ideas that would, if really used, lift the whole tone and temper of modern life. Unfortunately, many of these ideas are today buried under the jargon of technical scholarship and effectively insulated from contact with the common life.... Yet somehow we must thrust the results of research into the stream of common thought and action. The creative scholar is the hope of civilization, but his contribution does not become a social asset until it gets beyond the stage of inarticulate accuracy." And President Frank concludes that, to avoid disaster, "society must evolve some technique for using the results of scholarly research in the determination of its basic policies."[6]

This sound conclusion we should seriously take to heart. Unless the public mind can be put, and kept, in intimate and effective touch with science, our civilization will be a house divided against itself. There can be but one end to such a situation. Increasingly detached from everyday life, science will wither, like a plant whose roots are no longer embedded in the soil.

To avert so fatal a possibility should be a challenge to all believers in human progress, however they may differ from one another in the specific methods by which they think progress can best be furthered.

---

6. Glenn Frank, "The Outlook for Western Civilization," *The Century Magazine,* September 1925, pp. 628-629.

SPLIT IN THE CAMP OF PROGRESS

Chapter V

Split in the Camp of Progress

Some years ago I stood on Crown Point—that dizzy height of the Oregon Highway overlooking the Gorge of the Columbia, where the mighty river bursts its way through the Cascade Mountains to the sea.

It was a winter day. Storm winds were blowing in from the Pacific Ocean. Rebuffed by the towering Cascade mountain wall, the winds converged upon the Gorge, the one gap through the mountains to the eastern hinterland.

Far down the gorge, many miles away, I could see the storm clouds gather themselves together in vast billowing masses—then rush, roaring as through a giant funnel, covering me with spurts of rain or swirls of snow as they passed.

Yet, ever and anon, between the surge of those vaporous waves, rifts of blue sky appeared through which shafts of sunlight stabbed like molten swords, silvering the waters of the Columbia far below and glorifying the somber mountain walls, with their forest-clad slopes and glittering crowns of snow. Strange contrast of lowering gloom and radiant promise!

Thus may our world appear to thoughtful minds, gazing out over the striking panorama of modern life from the vantage point of the spirit. Storm cloud and sunshine, contrast and paradox, peril and promise: how will it all end?

There is a question that no man may surely answer; yet, that the query is momentous, none can deny. Indeed, the keynote of our age is becoming increasingly one of anxious interrogation. More and more, forward-looking men and women everywhere glimpse the loom of giant possibilities for both good and ill and are oppressed by the sense of impending alternatives of triumph or disaster. Even the less reflective masses grow dimly conscious that we live in crucial days that may determine man's destiny perhaps for centuries to come.

This combined challenge of fear and hope is quickening the race to a powerful ferment of thought and emotion. In every civilized country, active minds are pondering the problems of modern life and are feeling their way toward possible solutions. And since the civilized world is coming to be a closely knit intellectual unit, these diverse reflections and suggestions are constantly interchanged and spread to the uttermost ends of the earth, thereby stimulating fresh thought and further intellectual exchange.

Indeed, the extent of intellectual intercourse within the civilized world, together with the widespread appreciation of common difficulties, has led some idealistic minds to dream that, in the not-too-distant future, the basic problems of modern life will be dealt with by the concerted activity of the world's best brains—by the pooled intelligence of the planet, consciously working toward definite ends.

However premature this alluring vision of worldwide intellectual cooperation (perhaps culminating in a Supreme Council and Directory for Civilization) may be, it is certain that a much better understanding than now exists among forward-looking minds is possible, at least within the boundaries of present national and cultural areas. Every community possesses certain exceptionally influential individuals who formulate and guide public opinion. Taken collectively, these individuals constitute humanity's élite, and since most of them genuinely desire the progressive well-being of mankind, they form an élite of intelligence and goodwill.

Could a majority (or even a notable fraction) of this élite arrive at a genuine agreement on basic aims, together with a reasonable degree of cooperation in methods of attainment, the results might be momentous. The most striking successes in human history have been the work of able, resolute minorities. In our changing and uncertain age, what might not be accomplished by a powerful alliance of intelligence and idealism, assembled in a true Camp of Progress?

Unfortunately, no such alliance exists. Instead, we see the progressive elements split into many factions, often bitterly hostile to one another. Their potential power is thus frittered away, and the vast natural forces that we have ourselves unleashed are left to mold us in a blind and unregulated fashion. Today, as in the past, man continues to be the sport of the elements, notwithstanding his new knowledge and power of control.

Faced by the perilous consequences of such conditions, why is it that no concerted effort of human intelligence to remedy the situation has yet been made? The stock answer is that men differ too profoundly in their ideas as to what should be done and that we hold intellectual attitudes too diverse to make possible any effective understanding.

If this be true, the case is indeed hopeless. If even the wisest members of the race cannot study the facts of a situation, discuss them dispassionately, and arrive at a general understanding regarding courses of action, man may as well renounce the hope of guiding his destiny and must forever remain strictly subject to natural forces, like the other forms of animal life.

But is it true that intellectual understanding and effective action based thereon are impossible? If we look at the matter closely, may we not find that the source of the trouble lies not in the intellect but in the emotions; that what divides men irreconcilably is not rational thought but emotional processes?

In the chapter on "Our Unscientific Selves," we discovered that thought and feeling were extraordinarily intertwined; that conduct was induced mainly by our emotions; and that most of our so-called "thinking" was not deliberate, rational brain activity but more or less cleverly camouflaged emotional reactions. Furthermore, we learned that these mental shortcomings were common to all of us; that even the leaders of the race had in their make-up the same irrational traits as their less distinguished fellows, the chief difference being that most well-endowed individuals possessed an intellectual grasp and emotional control capable of clearer thought and wiser action than the mass of mankind.

Our new knowledge about ourselves throws a strange light on many of the "intellectual" differences over which we quarrel. In that new light we observe that most of our "principles" and "convictions" (especially if they be "self-evident") are mere "rationalizations" of beliefs and prejudices and are thus rooted in the emotions, however luxuriantly they may flower in the intellectual air. Even the loftiest minds are ferments of mingled intellect and emotion, with rational thought distilling itself slowly and sparsely through the cooling medium of conscious endeavor. With the mass of mankind, rational thought is rare and is easily overwhelmed by floods of primitive passion welling up from the emotional depths.

We are now able to analyze the true nature of opinion. However pronounced the intellectual differences that may arise, it is not these which are primarily responsible for the embitterment which so often ensues. What turns discussion into controversy and what creates dogmatism, partisanship, hatred, and intolerance are, at bottom, the emotional reactions aroused by the impact of distasteful or disturbing ideas. That it is not the ideas themselves but the emotional reflexes that make the trouble is easily shown. Do we not all remember some occasion on which we encountered a new idea that did not perceptibly affect our beliefs or interests? In that case, not only was there no defensive emotional reaction, but we probably experienced a sensation of pleasurable curiosity. Our minds opened, our imagination got to work, and we found ourselves playing with novel possibilities. Indeed, we can doubtless recall other occasions when we encountered ideas that, though unpleasing, were presented so tactfully and on so high an intellectual level that we succeeded in overcoming our emotional reactions and considered the idea with a fairly dispassionate and open mind.

These simple instances illustrate the much more complex and intense conflicts between intellect and emotion that occur in connection with every important question of both public and private life. How momentous their consequences may be will be seen in later chapters, where we shall survey contrasting attitudes and note the bitter quarrel between "Heart" and "Head," which rages around so many problems of our day. The point for us to note here is that the manner in which a problem is faced depends largely on the emotional attitudes involved. In other words, from this angle, it does not so much matter *what* men believe as *how* they believe.

Again, we may as well recognize that with most men the mainspring of action is impulse and prejudice rather than rational thought. Only a small minority are capable of thinking clearly and constructively when emotions are aroused. For this reason, we can clearly foresee that, so far as the masses are concerned, heart will speak louder than head, and "burning issues" will be popularly treated on an emotional rather than a rational plane.

Furthermore, the latent power of the emotions in even the loftiest minds makes it likely that the partisan bitterness and intolerant dogmatism engendered in moments of emotional crisis will infect, by crowd contagion, even the intellectual élite. That is why, at such moments, the wisest leaders are apt to lose their clearer vision and descend to the passions and prejudices of their followers. A good instance of this is in wartime, when the coolest and best-informed minds seem stricken with intellectual blindness and share the hysterical fears and delusions of their more ignorant and impressionable fellow citizens.

Our analysis of the factors determining opinion and conduct might at first sight seem to lead us to rather pessimistic conclusions, especially to a disparagement of the emotions and a corresponding exaltation of the intellectual faculties. That, indeed, was the tendency among thinkers in past times, when intellect and emotion were believed to be separate and sharply contrasted things. Philosophers acclaimed the superiority of intellect and even went so far as to depict a quality termed "pure reason."

Our new knowledge about ourselves has, however, sent all such philosophical lucubrations into the discard. We now know that there is no such thing as "pure reason," that intellect and emotion are not separate but inextricably intertwined, and that both are intimately linked with the other bodily processes, so that a man suffering from chronic dyspepsia can hardly avoid being a confirmed pessimist, whereas one with a good stomach and active glands will usually see the world in roseate hues. If in our analysis we have seemed to isolate the intellectual and emotional factors, it has been merely in order to comprehend certain aspects of an exceedingly complex and subtle process—the functioning of human personality.

Viewed in their true relation, there is no real antagonism between intellect and emotion, which are both vital parts of an organic whole. A French wit once exclaimed: "Women are the devil; one can live neither with them nor without them!" With less cynicism, we might apply this to the emotions. Our emotions may often vex us and may sometimes get us into serious trouble; yet they are the mainspring of effective action and the source of inspiration. If "pure reason" could exist, it would probably be about the last thing that we should want, for intellect wholly untouched by emotion would be inhuman, presumably more diabolic than divine.

Let us, then, realize that a harmonious combination of intellect and emotion is needed, alike by the individual and the race. We adjust ourselves to our world and our fellows, not merely by thinking but also by acting, and before thought can be translated into action, the emotions must come into play. Our aim should, therefore, be to discover the respective functions of intellect and emotion and, on the basis of that knowledge, to deal with the problems of life. That is what we must do as individuals if we are to be successful and happy, and that is what we must do collectively if society is to attain prosperity and progress.

It is our failure to appreciate these wholesome truths that is responsible for many (perhaps most) of our present perplexities. We have among us a host of intelligent and idealistic men and women who earnestly desire the progressive well-being of mankind. Their potential power, if effectively coordinated, would produce almost incalculable results. Why does this not happen? Why, on the contrary, do we see them split into bitterly hostile factions, working mainly at cross-purposes? They all have the same basic vision: human progress. Why, then, does not this kindred vision yield a common sympathy and understanding, forging them into an effective alliance of idealism and knowledge for the furthering of their larger aims, despite differences of opinion as to how the goal may best be attained?

The answer to this query lies ultimately in our failure to distinguish between intellectual and emotional differences. In preceding pages, we analyzed this point and found that what divides men irreconcilably is not thought but feeling, not intellectual concepts but emotional processes. Yet nearly everyone believes that it is the *ideas* that make the trouble! Accordingly, most people strive persistently to formulate some intellectual "principle" or frame some abstract "program" that will persuade and reconcile.

Vain delusion! The only way to form a true Camp of Progress, flying a standard to which the wise and righteous may repair, is to cultivate an attitude of mind, *and even more a state of feeling,* by which intelligent understanding and emotional comprehension will become mutually possible among those genuinely desiring human progress, however diverse may be the methods by which they conceive that such progress should be furthered.

And right here there once more emerges that *Spirit of Science,* which we have already seen to be the inspiration of our time. How can we cultivate the intelligent understanding and emotional comprehension that we seek, except by the scientific attitude of mind? Here we have the method by which thought and feeling can be harmonized and brought into effective collaboration. No matter what the problem, behind it lie certain *facts*, which have been, or can be, scientifically ascertained. There, at least, is a task in which loyal cooperation is possible! However else they may differ, sincere and thinking men can get together in an honest search for the facts and an objective analysis of them after they have been found.

This does not mean that everybody would agree. It might well be that the searchers would so vary in their intellectual and temperamental make-up that they would interpret the facts in widely different fashion.

Yet that does not invalidate the scientific method. On the contrary, it makes it all the stronger. In considering our problems, what we should be after is not a unanimous vote, but *the truth!* If the facts are so complex or so apparently contradictory that various interpretations are possible, the sooner we find it out, the better for us. In fact, the shock of different intellectual viewpoints is one of the chief aids to progress, by stimulating thought and quickening fresh research for new facts at one and the same time.

Yet, however marked may be the differences of ideas and however keen the discussions that ensue, these *need* not provoke hatred or intolerance if they are kept on a truly intellectual plane. Emotion emits merely heat without light. On the other hand, rational thought should produce light with, at most, a gentle warmth that does no harm. And the scientific method is a supremely efficient cooling system, through its resolute facing of facts and its insistent search for objective truth—thereby freeing the mind from those smoldering fires of passion and prejudice which, despite the best intentions, tend to make men "believe what they wish to believe."

How the genuine application of scientific methods to our everyday problems would raise their treatment to a higher level and would immensely facilitate their solution can be realized by a few homely illustrations. Suppose, for instance, that a legislative body debating a bill, or a citizens' committee considering a pet reform proposal, should conduct their proceedings in the same spirit and by the same methods as a group of scientists engaged in an experiment. Do we realize what their first act would be? Well, before even venturing to discuss the matter in hand, the first thing the legislature or the committee would do would be to calculate what scientists term their "degree of probable error." In other words, they would try to determine as far as possible the amount of intellectual mistake or emotional bias that might unwittingly exist in their minds and thus influence their conclusions! Can we picture to ourselves, without ironic laughter, the spectacle of politicians or "reformers" calculating their "degree of probable error" at *any stage* of their proceedings? Yet this is just why our political and social "thinking" is today so warped by prejudice and so barren of constructive ideas.

When our leaders begin consciously to cultivate a mental attitude in which political and social problems are viewed objectively, with cool weighing of facts and far-sighted attention to probable consequences, we shall see social policies determined in a truly progressive spirit, instead of by either a blind clinging to traditional prejudice or rash appeals to sentiment and high-sounding phrases that ignore or defy reality.

And when our leaders thus truly lead, they will be astonished at the response they will get from their followers. It is doubtless true that only a small minority are capable of constructive thought, but it is also true that the masses can learn to appreciate the value of its results. If one-tenth of the time and effort now spent on emotional appeal and partisan propaganda were devoted to impressing upon the public the practical importance of the scientific attitude and method in everyday life, their significance would permeate the popular consciousness and would subtly influence the general trend of the age. We should no longer witness the bitter conflicts that now rage between emotion and reason, between Heart and Head; because, though most men might still be moved by the heart, they would listen to the head before proceeding to definite action.

Then, indeed, would the Spirit of Science inspire us all. For, as Wiggam has finally said, "Only genius can create science, but the humblest man can be taught its spirit. He can learn to face truth."

How far removed we today are from such a happy state, and how grievously we need to work toward such a goal, can be seen by a survey of current problems, which we will now essay.

HEART AND HEAD QUARREL

Chapter VI

Heart and Head Quarrel

The quarrel between Heart and Head is not only a very serious but also a very pathetic aspect of our time. Here we are in our strange new world, faced by problems and perplexities of unprecedented magnitude. Your mythical Visitor from Mars might suppose that so grave a situation would inspire the Sons of Earth to the harmonious exercise of all their faculties. Alas! The Visitor from that older (and presumably wiser) planet would not realize the shortcomings and perversities of human nature. To his profound amazement, he would soon discover that on many. of the most vital questions of the day, intellect and emotion were ranged on opposite sides; instead of working together for common ends, thought and feeling were wasting their energies in mutual recrimination. What the Celestial Visitor would think when he observed men of real idealism and intelligence, the very flower of the race, belaboring each other instead of combining in the task of enlightening and guiding their less gifted fellows may be left to a Martian imagination.

How crying is the need for such a task, a moment's reflection will show. Despite the vast extensions of knowledge and power *that a few superior minds have conferred upon the civilized portion of mankind,* the amount of sheer ignorance, superstition, and general backwardness persisting is literally appalling. As one writer well says, "It is in his possibilities rather than in his actual state of development that man is a superior being. Those possibilities have to some extent been realized in the educated few. When we speak of civilization, of modern life, and of scientific advance, we fix our attention exclusively on the educated *élite*. In asserting, for instance, that nowadays nobody doubts that the earth is round, we are referring to a very small minority of the human race, composed of perhaps half the adults of Europe, America, and Australia, with a small sprinkling of Africans and Asiatics; in all, perhaps ten percent of the total population. If a vote were taken of every man, woman, and child in the world to decide whether the earth is round or flat, there would be a majority of perhaps 9 to 1 in favor of a flat earth."[1]

This astounding contrast between the enlightened few and the unenlightened many makes all sorts of trouble. In the first place, it creates a profound disharmony

1. E. E. Fournier d'Albe, *Quo Vadimus? Some Glimpses of the Future,* pp. 16-17 (New York, 1925).

between our world and ourselves. Modern science is literally bombarding us with new conditions—and new ideas. But these novelties encounter minds mainly "set" in molds of traditional belief and opinion handed down from pre-scientific times and usually anti-scientific in character. Ill-adjusted as we thus are to receive them, the new ideas shock and disturb us, thereby rousing emotional reactions that either close the mind to the new ideas or else admit them in a muddled, distorted guise. And since our beliefs are deep-seated, their emotional resistance is powerful and hard to overcome.

Indeed, this widespread resistance produces a sort of emotional fog, which not only pervades the masses but also obscures the vision of many superior minds. This is especially true of idealistic natures, in whom a good intelligence is often overborne by strong sympathies and enthusiasms. In such cases the intellect becomes the slave of the emotions, the upshot being the utterance of arguments that *seem* to be the product of deep thought, but which are really nothing but "rationalized" feeling. Many of our radical "intellectuals" are of this type, and the same is true of certain religious spokesmen. The "Parlor Pink" and the "learned" Fundamentalist are emotional brethren, however much they may differ in other ways.

The trouble is that these "rationalizations" are often so persuasive, and the rationalizers themselves are so sincere, that multitudes of persons are led to adopt ideas that run counter to the real world and to human nature, and which may have the most disastrous consequences.

The amount of disguised emotion today masquerading as thought is literally astounding. Nowhere do we see it more plainly than in the current tendency to "think in phrases"; in other words, to *avoid thinking* by embracing some clever catchword or resounding "principle" that *seems* to settle the question and bring conviction to perplexed souls. This is a very insidious evil, because it so subtly debauches the intellect and kills creative thought. Constructive thinking is hard work, while nothing is more painful than a disturbing idea. How much simpler to take a mental shortcut, which we fondly imagine will do the work and ease the pain!

On this unfortunate trend of the times, Wiggam speaks so pungently that I cannot do better than quote him. Says Wiggam: "This is a sloganized age, an age of searching, not for solutions of social problems, but for what Professor James calls 'solving words.' Democracy, progress, Brotherhood, Communism, Uplift, and Humanity are not solutions for anything but mere solving words.... Just because a thing is democratic or progressive without any reference to where it may progress toward, it must be right. It has exactly the right *name*. As James points out, Solomon could control the evil spirits because he knew the right names of all of them. Address an evil spirit by the right name and you've got him. And this age is obsessed with the idea that social evils will yield to the same treatment.

"If a 'democratic' remedy fails to cure anything, it is proof, not that it is the wrong remedy, but that it is not democratic enough. Pour in a little more 'democracy'! To calculate, to measure, and to analyze the psychology of human motives; to add up columns of figures; and to calculate standard deviations and coefficients of correlation; this requires hard work and intelligence. It requires intellectual men. It requires men who want to solve things instead of finding solving words for them....

"But the faith in solving words in the place of hard-won solutions reigns supreme over this age. There were never so many problems, so many solving words, nor so many people who believed in them. Yet they never have solved anything. Nothing but intelligence and goodwill, usually extended over long periods of time, ever solved any social problem."[2]

Such substitutes for thought, bandied about in argument or hurled in bitter invective, form the bulk of popular discussion of present-day affairs. No wonder public opinion is mainly on the emotional plane! No wonder that the great field of politics, for instance, is dominated by traditional methods, almost uninfluenced by objective, dispassionate standards. Yet nothing but the application of scientific methods can rescue politics from its present muddling inefficiency. And, in the last analysis, the way to bring this about is by the spread of the scientific spirit and attitude in the public mind. The progressive liberalism and open-mindedness of the scientific spirit are absolutely necessary for a people if it is to succeed in truly ruling itself through rational public opinion. Yet today the public seems actually afraid of science in politics, preferring to trust the "professional" politicians who play the game according to the old rules with the old results!

One of the most hopeful experiments foreshadowing the introduction of scientific methods into the political field is the establishment of the National Conference on the Science of Politics, which for a number of years has held annual sessions at which this great question is studied and discussed. The Association is broad-based, its membership being drawn from nearly all the political and social sciences, as well as from men in active political and business life. Its aims and methods were ably stated by its president, Dr. Arnold B. Hall, during the Conference of 1925.[3]

Asserting that, politically speaking, we are still "jungle-folk," Professor Hall went on to say that only the application of scientific methods could "take the hunch out of politics," eliminating the factors of ignorance and prejudice, just as science has done for medicine and as it is doing for business and industry. "Businessmen," said he, "spend millions applying the methods of science to the study of human behavior in advertising, in vocational selection, and in mental measurement. But nothing is done to study human behavior in relation to politics. Yet what is of most value to society: knowledge of the most effective way to sell laundry soap or knowledge of how society reacts to political situations? We put our best thought into the selling of soap, but the controlling principle for politics is chance, haphazard experience, or the more or less skillful guess of the practical politician. ...

"We must find the elements and principles of social control, just as we have found the principles of business control. Advertising, for instance, has a claim to be a science because, if you go to a competent agency, they can tell you in the majority of cases just what will be the reaction to any given line of advertising. That is what we call predictability. There will be no political science until we have

2. A. E. Wiggam, *The New Decalogue of Science*, pp. 190-192.

3. Special statement made during the Conference held at Columbia University, New York City, to a representative of *The New York Times* and published in its issue of October 11th, 1925.

studied principles and causes deeply enough to predict with at least approximate certainty what will follow from a certain line of political conduct." And Dr. Hall concluded by saying that the way to attain this is through the methods employed in any other branch of science; namely, by scientific observation, experiment, objective analysis, and statistical treatment of fact-data.[4]

However, though activities like those above-described may foreshadow a better future, we must not forget that at present they are the work of a very small minority, with almost no popular backing. Indeed, attempts to apply scientific methods to practical matters often encounter not merely popular distrust but also the bitter hostility of many persons of more than average intelligence. It is precisely these conflicts within the ranks of the intellectual élite, these "splits in the Camp of Progress," that do so much to nullify the potential influence of the forces of intelligence and idealism. In the preceding chapter we analyzed the reasons for these conflicts and found them due to emotional rather than intellectual differences.

Can we discern any general line of cleavage running athwart these controversies? I think we can. I believe we can note a fairly definite distinction in attitude and outlook between those who recognize the broad implications of science and incorporate its spirit into their effective philosophy of life and, on the other hand, those who have not really assimilated modern science into their thought and action. What our age needs, perhaps above everything else, is a genuine and general assimilation into its idealistic and cultural scheme of the new facts, ideas, and methods discovered by modern science. Later on, we shall see that the way to do this effectively and harmoniously is through an intellectual and emotional development, which we may term *Scientific Humanism.* With this suggestion in mind, let us continue our survey of current problems, noting the difficulties caused by the lack of such a humanistic attitude.

In our analysis of man's nature, we saw how hostile the emotions are to ideas that disturb beliefs and prejudices. That is only another way of saying that emotionally we prize fixity and certitude. We like to have things that really "matter" to us "settled"—if possible "once for all," and we instinctively welcome "principles" that convince us and set our troubled minds at rest. Of course, that is only one side of our nature. We have other faculties, like curiosity and critical analysis, which break the molds of tradition and impel us to change. Yet a fresh mold quickly forms. An idea, once really accepted, is soon transmuted into a *conviction*, emotionally defended perhaps more jealously than beliefs traditionally held. This is well illustrated in the field of religion, where the convert is, notoriously, more ardent than those born in the faith.

In a wise balance between the factors of conservatism and innovation, thus avoiding the extremes of both dull stagnation and chaotic change, lies the secret of human progress. As a general truth, this has long been known. For instance, an

4. The statistical handling of data is a branch of scientific activity whose importance is especially unappreciated by the general public. One of the outstanding leaders in this field is Dr. Frederick Adams Woods, whose statistical studies have yielded the most enlightening results in branches of knowledge as far apart as biology, history, and finance. Some of his best-known studies are *The Influence of Monarchs, Heredity and Royalty,* and *Is War Diminishing?*

excellent formula of social progress is contained in St. Paul's admonition: "Prove all things; hold fast that which is good."[5] The difficulty lies in applying the general principle to particular times and circumstances. We have seen that the basic problem of our age is adjustment to the extraordinary new conditions that modern science has produced. We have seen that a necessary element in the process of adjustment is the harmonious cooperation of our intellectual and emotional faculties. Lastly, we have seen that such cooperation is dependent on our adoption of the scientific spirit and attitude. Let us now see how this spirit and attitude can be effectively translated into *action* in our everyday affairs.

At first sight certain difficulties might seem to arise. The research scientist is so distrustful of his own as well as of his fellows' intellectual and emotional limitations that, however extensive his store of fact data, he hesitates to draw definite conclusions. Above all, he fears becoming dogmatic, thereby losing his open-mindedness and objective vision. However boldly he may experiment in his laboratory, he shrinks from applying his tentative conclusions (his "working hypotheses," as he calls them) to the everyday world of men and affairs. He therefore logically tends to suspend judgments and adopt an attitude of "wait and see," and he instinctively dislikes having his findings escape from his laboratory control into the outer world, where they must inevitably become both fixed and limited by practical necessity.

And yet, somehow or other, the findings of science must be applied to society and put to practical use as soon as possible if our age is to avoid disaster. Here are all sorts of vital problems clamoring for solutions. They cannot be wisely dealt with save in the light of scientific knowledge. Admitting that in many cases science has as yet comparatively little to say, let us nevertheless hear that little and then act upon it to the best of our ability. We can do this if we act in the scientific spirit, ready, like the scientist in his laboratory, to keep an open mind and to change our line of action if fresh evidence indicates that we are on the wrong track.

How regrettably slight the influence of the scientific spirit today can be seen by the way many attempts to put scientific knowledge to practical use are literally bedeviled by emotionalism. Wherever they run counter to established beliefs and prejudices, such attempts rouse emotional storms that often turn what should be fruitful intellectual discussion into barren dogmatic controversy. The best examples of this are the quarrels waged over the practical application of scientific findings concerning human nature. During the past two decades, biology, psychology, and other sciences have amassed a great store of data, which amazingly modifies traditional ideas about ourselves. The potential significance of these discoveries is so striking that attempts have been made to apply them to social policy. Accordingly, we see numerous studies on "eugenics" or race betterment, on the relation of heredity to environment, on the measurement of intelligence, and on kindred matters.

Now here are golden opportunities to apply science to everyday affairs. If these findings are true, they are obviously of tremendous practical importance. And surely it is equally obvious that their truth depends not upon our wishes but upon *fact*. However much we may like or dislike them, if they are so, they are so—and

5. I Thessalonians: 5:21.

the sooner we quit hugging illusions and recognize realities, *the better for us.*

Furthermore, we should often translate scientific findings into practical policies, even where discovery is admittedly in the tentative stage. If, for example, the weight of evidence indicates that intelligence is inherited, and if scientific tests are devised that apparently measure inborn intelligence with a fair degree of accuracy, we should apply these tests to our educational system; we simply cannot afford to continue the inefficiency and injustice of present educational methods for twenty or thirty years until perfected tests may be thoroughly worked out. Again, if biological research tends to show that the quality of the race is rapidly deteriorating, we simply must take active steps to check the breeding of low-grade stocks and encourage the increase of high-grade strains, because if we do nothing but let biologists accumulate fresh data, it may turn out to be a case of locking the door after the horse has been stolen. Absolute proof is the correct ideal for the laboratory, but "balance of proof" is the proper rule for everyday life.

Offhand, one might think that the application of science to the vital problems of our age would inspire a high idealism that would spur the forces of intelligence and goodwill to concerted, harmonious endeavor. To view objectively, to suppress prejudice, and through rational discussion to arrive at constructive social policies might seem to be the course of conduct that would normally suggest itself to all those aspiring to further human well-being. Far from it! Instead of harmony, we see the Camp of Progress split wide open on the most vital questions of the day. About matters like heredity and race betterment, what should have been intelligent discussion has degenerated into emotional controversy, with the bitterest notes struck by persons who especially pride themselves on their "progressivism"!

The unseemly wrangle over the intelligence tests is a good case in point. The formulators and advocates of the tests were at once assailed by a clamor of protest, clearly emotional in the main. The trouble was that the scientific generalization behind the tests (the concept that men differ widely and ineradicably in inborn intellectual capacity) ran counter to the older idea that such differences either did not exist or could be nullified by education and other environmental influences. The shock to this established idea instantly aroused defensive emotional reactions, which were not controlled but were given free rein. The upshot was that what might have been fruitful criticism degenerated, for the most part, into mere "rationalizing" or even into downright personal abuse. Not only were the tests themselves impugned, but the framers and advocates of the tests were bitterly assailed as "enemies of democracy" and creators of a "new snobbery"—in short, as sinister and generally undesirable persons.[6] All of which merely showed that the intelligence testers had trodden on somebody's emotional corns.

Now this sort of thing may relieve the feelings and may arouse popular

6. The literature of this controversy is voluminous. A representative cross-section of it is contained in the files of *The New Republic* from the autumn of 1922 through the first half of 1923. Herein will be found a number of articles by Walter Lippmann and others, assailing the tests and their sponsors, together with replies from Professor Lewis M. Terman, a leading specialist on the tests, Professor William McDougall, and others.

sympathy—and prejudice. But it is a costly luxury because it seriously impedes social progress. Let us note the consequences of the wrangle over the intelligence tests—remembering, of course, that it is typical of many similar controversies.

In the first place, the emotional attitude of the assailants prevented them from viewing the tests fairly. Filled with apprehension as to how the tests *might* be misused, the critics were largely estopped,* not only from objective analysis of what the tests actually were, but also from seeing how they could be usefully employed. Their attitude being thus intellectually negative, their criticisms were obstructive or destructive rather than constructive in character.

* Note: In simple words, estoppel is a legal rule that stops someone from going back on their word or denying something they previously said or did, especially if someone else relied on it and would be harmed if they changed their story. It's about fairness, preventing people from unfairly contradicting their past actions or statements to gain an advantage or cause another person loss.

On the framers and supporters of the tests, likewise, the controversy produced regrettable effects. Realizing that criticism of the tests was mainly emotional, they were annoyed by its irrationality and were disinclined to take it or the critics seriously. In other words, the defenders also experienced beclouding emotional reactions, which tended to harden their opinions and close their minds.

The upshot was that discussion sank steadily from the intellectual to the emotional plane, thereby preventing what might have been a thoroughgoing consideration of the subject by all the intelligent elements of the community. The formulation of an enlightened, well-rounded conclusion, and its translation into effective social action, were thus distinctly retarded.

The most unfortunate feature of this and kindred controversies, however, remains to be told. It is the emotional appeal to the general public and the arousing of uninformed and ignorant prejudice that ensues. In our democratic age this is becoming an increasingly serious matter. Formerly, "public opinion," as regards intellectual novelties, meant a small, educated minority. The illiterate masses did not count because they simply were not aware that anything was going on. Today the situation has been radically altered. In the leading nations of the civilized world, universal education and a ubiquitous press have made the term "public opinion" mean the bulk of the population. But an enormous *literate* public does not imply an *educated* public. It is more apt to mean a public able to read but unable to appreciate what it reads.

Furthermore, dominated as most men are by their feelings, it means a public much more susceptible to emotional appeal than was the smaller but better-informed public of earlier times. Therefore, when large sections of the educated elite give way to their feelings, abandon their role of intellectual stabilizers, and call the popular masses into the controversy, they are starting something that may end as surprisingly to themselves as to their opponents. For if great multitudes practically untouched by the scientific spirit, ignorant of the facts, and unable to distinguish fact from fancy are emotionally aroused, there is no telling what may ensue. Once persuaded that the matter vitally concerns them, they may, in their ignorance and prejudice, come to the most foolish and dangerous conclusions.

Yet no matter how disastrous to social progress the conclusions may be, those aroused multitudes are able to make their power felt. How many projects generally recognized by informed opinion as scientifically sound and socially needed are today incapable of realization because they are not "practical politics"—in other words, because they have against them a latent or active popular hostility! In fact, how many scientific advances already made may be undone by those same irrational forces of popular ignorance and prejudice! The "Fundamentalist" campaign against the scientific theory of evolution is merely the most dramatic instance of a widespread trend of the times. Indeed, it is over the relations of science and religion that clashes between reason and emotion are most bitter and intense. To this supreme phase of the quarrel between Heart and Head, let us now turn.

SCIENCE AND RELIGION

Chapter VII

Science and Religion

The conflict between science and religion is the most distressing problem of our age. That the highest activities of the human mind and the loftiest aspirations of the human spirit should appear in many ways opposed; that men of real intelligence and idealism should be thus ranged in hostile camps, instead of leagued in an alliance of wisdom and righteousness: here, indeed, is tragedy for earnest souls! How comes it? and, Whither is it tending? are queries that must spring instinctively to the lips of every sincere lover of mankind.

At first sight the wave of religious hostility to many of the findings of modern science, which has arisen of late years and which is most dramatically manifest in America by the Fundamentalist movement and its campaign against the theory of evolution, might seem to be a mere revival oi the controversy that took place half a century and more ago over the publication of Darwin's *Origin of Species*. At that time public opinion was passionately divided along religious and scientific lines. England was then a prototype of Dayton, Tennessee. From pulpit and platform Darwin was denounced as a sort of anti-Christ. Pious folk spoke his name with bated breath, and many persons dared not admit that they had read the book they criticized. The controversy culminated in the Homeric duel between Huxley and Gladstone, the latter's polemic on "The Impregnable Rock of Holy Scripture" reading almost like a speech of Mr. Bryan's.

And yet, despite its apparent identity, the anti-scientific religious trend of today differs from its predecessor in one important respect—its much more "popular" character. The "public opinion" that bestirred itself over Darwin half a century ago meant little more than the educated minority. The bulk of the population was not concerned. Knowing and caring little about what was going on in the "upper circles," the masses, undisturbed in their traditional beliefs, went about their own affairs. If Darwin, Huxley, and the other mid-Victorian champions of evolution meant anything at all to the common man, it was merely as a set of impious atheists foredoomed to failure in their rash attacks on the strongholds of dogma and tradition.

Very different is the situation today. In the sixty-seven years that have elapsed since the publication of *The Origin of Species,* momentous changes have taken place. The spread of popular instruction has created an articulate "public opinion" that embraces the bulk of the population, while the correlative spread of democratic institutions has endowed the masses with the vote, thus enabling them to translate their opinions into political expression and action.

In the preceding chapter, we noted the unfortunate consequences that might ensue from the uninformed and emotional decisions of mass opinion on political and social policies. This is even truer where religion is involved, because in such cases passion and prejudice are so easily aroused. Here again we observe the grave disharmonies of modern life. The past two generations have witnessed a growing disproportion between the material applications of science and the appreciation of the scientific spirit and a widening intellectual gap between the educated élite and the uneducated or half-educated masses.

Of this trend, America is perhaps the most striking example. Materially and mechanically, the American people are very advanced; intellectually and philosophically, they are not. Of course, America possesses an intellectual elite equal to any in the world, but large elements of its population are decidedly backward in outlook. This accounts for the extraordinary contrasts of American life. Hence arises the curious situation of people literally immersed in "up-to-date" comforts and conveniences, using mechanical marvels like telephones and motor cars as actual necessities of life—and yet so untouched by the scientific spirit and attitude that have made those inventions possible that their beliefs remain basically unadjusted to our scientific age. No wonder such people react violently to suggestions that the fundamentals of traditional faith are being undermined and try to use their political power to safeguard religious doctrines to which they fervently cling!

There is the root cause of the Fundamentalist movement: a deep popular distrust of the scientific spirit and attitude. Of course, it is illogical and inconsistent to condemn that spirit and yet welcome its fruits. What an absurd paradox is the spectacle of a Fundamentalist preacher "broadcasting" sermons on the wickedness of science! To be true to his convictions, he should emulate the Oriental mystics who denounce the whole development of Western civilization as essentially evil. Gandhi, for example, condemns all our scientific discoveries, from cotton machinery to preventive medicine. Our Fundamentalist divine should, therefore, logically eschew the radio as a device of Satan.

But, of course, logic and consistency have little to do with the matter. Here are millions of people today confronting the first shock of disturbing new ideas. They are facing the same necessity for readjustment that the educated elite went through half a century ago. That readjustment was, on the whole, successfully made, so that most educated, forward-looking men and women have assimilated the broad conclusions of science into their philosophy of life. To them, therefore, the current campaign against evolution seems as amazingly incomprehensible as would be a widespread propaganda in favor of a flat earth. Our intellectuals forget that what has become self-evident to them is shocking and repugnant to multitudes just aware of the idea and therefore only now able to voice their protest. The fact is that the spread of the scientific spirit and the adjustment of the public mind to scientific knowledge have been vastly overestimated. The Fundamentalist movement makes us, for the first time, really aware of the deep intellectual and spiritual gulfs that today divide the enlightened élite from the unenlightened majority.

And this is by no means an exclusively American matter. On the contrary, it is an issue that every civilized nation must sooner or later confront. America's present controversies over science and religion have been watched by Europe with amused incredulity. Fundamentalist activities have called forth satirical condolence for the *Benighted States of America,* while the Tennessee "Monkey Trial" made the whole civilized world rock with laughter.

And yet, our European friends might perchance have profitably considered Dayton as a portent rather than as an idle jest with no possible meaning for themselves. Doesn't Europe also have its unenlightened masses? Today they are not merely intellectually backward but also so accustomed to the cultural leadership of their intellectual betters that they hardly figure in articulate public opinion. But how will it be tomorrow, when the implications of science have been driven home to them, when popular education has overcome their cultural timidity, and when religious reactionaries urge them to defend their traditional beliefs by means of the political power that universal suffrage has put into their hands? What, for instance, will the vast peasant masses of Eastern Europe do if they are ever deeply stirred by anti-scientific religious feeling? Already certain events that have occurred in central and eastern Europe during the past few years seem to point rather ominously in that direction. Some of the peasant movements that have arisen there disclose a distinctly fanatical trend. In Croatia, especially, the peasant leaders exhorted like modern Savonarolas, denouncing the whole fabric of our civilization. In Russia the peasants harbor similar feelings. Not long ago Maxim Gorky stated that the Russian muzhik profoundly distrusts our way of life. Modern "progress" does not appeal to him; the intellectuals and their works are regarded with deep suspicion. Gorky relates how, after addressing a peasant audience on the subject of science and the marvels of technical inventions, he was criticized by a peasant spokesman in the following manner: "Yes, yes, we are taught to fly in the air like birds and to swim in the water like the fishes, but how to live on the earth we don't know." In Gorky's opinion Russia's future lies in peasant hands. Ultimately, they will be Russia's masters. But Gorky is far from easy as to the character of that mastery because he does not believe that peasant hostility to modernity will abate. Should that be the case, peasant rule in Russia would be hostile to the very spirit of modern civilization.

Such is the issue before the modern world: the enlightenment of the unenlightened majority and the reconciling of alarmed emotions to scientific knowledge. For the educated minority who have already made this adjustment, the prospect is not alluring. To fight once more the old battles that they imagined were over and done with half a century ago seems at first sight a dreary and irritating task.

Yet the task is clear, while its vital importance provides at once a challenge and an inspiration. When we consider what science and religion both mean to mankind, we must recognize that anything like a lasting feud between those outstanding intellectual and spiritual forces would render impossible the harmonious development of our civilization. Surely, our new knowledge of human nature teaches us (as we have already seen in these pages) that intellect and emotion are not basically opposed but can be truly coordinated. The reconciling of science and religion is as possible for the masses as it has been for the classes, if approached in the right way.

And that way lies not through gibes and jeers, which merely provoke fresh bitterness and hostility, but through sympathetic insight that can pierce to the heart of the trouble. Here, more than anywhere else, the elite of intelligence and goodwill can play their stabilizing, comprehending, and reconciling role. Not until the common man realizes that science will not rob him of his spiritual aspirations but that, on the contrary, science can give him a broader outlook, a fuller and richer life, a sounder morality, and a deeper contact with the ultimate mysteries of the universe will his fears be allayed and his heart be opened to let the spirit of science in. Then, and then only, will the great schism of modern life be healed.

Surely, there can be no basic antagonism between man's highest mental and spiritual aspirations, between his fearless quest for objective truth and his yearning quest for contact with the Ultimate! These exalted human endeavors are not sundered or opposed; they are twin aspects of our supreme task: the harmonious adjustment of Man to the Universe and the reverent communion of the finite with the Infinite. There is the core, alike of true science and of true religion! Strip away the obscuring husks of hypothesis and dogma, and you will find the same bright kernel within.

Does not the trend of the best scientific and religious thought today bear witness to this? They are obviously converging in attitude, getting closer together in mutual understanding. We have already seen how Nineteenth-Century materialism has vanished in a whirl of protons and electrons, with Matter, Space, and Time blending into mere phases of an underlying Reality. And, just as modern science has come to recognize the essential spirituality of the universe, so liberal religion has come to recognize the scientific spirit as a vital part of its reaching out after God. This current trend in religion is well expressed by a liberal churchman in the following lines: "If one may venture briefly to state the ruling tendencies of religious experience today, it would be after this manner. First, a new realism of faith & reverent and fearless demand for reality, at once urgent and austere and impartial, the desire to found faith upon fact, on the ground that it is the truth alone that can set us free from error and illusion. No institution, no document, and no dogma is exempt from its scrutiny. Second, a desire to live spiritually and think scientifically, uniting the old values of the spirit with the new vision of the world and its laws, which means that the spirit of science is itself religious in its veracity, its humility, and its single-hearted devotion to the truth."[1]

Surely, in the last analysis, the spirit of science and religion is one. For, as

1. Rev. Joseph Fort Newton, "The Changing Vision in Religious Ideals," *International Book Review*, April 1924, p. 359.

another religious liberal remarks, "The passion for veracity, the adventurous and unmercenary love of truth, the faith that more light is yet to break, devotion to God's ways not only made known but to be made known, the high impersonality which must correct all arrogant and inadequate individualism, the catholicity of scientific knowledge as against provincial sectarianism . . . all these are indubitable signs of religion, and are to be found in modern science perhaps more clearly than in modern ecclesiasticism.[2] How close and sympathetic are the present attitudes of science and liberal religion? They can be gauged by this judgment of a representative scientist in the same connection: "We are today in the midst of a religious revolution that is going on so quietly that many do not notice it, although it is a greater and more deep-going revolution than any since the early years of the Christian era. We are witnessing great changes in the attitude of the churches on questions of faith and science. The spirit of science has entered into religion. This spirit demands not uniformity of belief but uniformity of aim, not absolute and perfect truth but the best available truth, not authority but evidence, not words but works; and more and more, religion is demanding these things."[3]

Indeed, one of the clearest and most farsighted leaders of religious thought glimpses nothing less than a close alliance of science, religion, and philosophy, united in a common cause—the exploration of the cosmos and the harmonious adjustment of man to the varied aspects of reality. Says L. P. Jacks: "Instead of regarding these three as rival claimants for the explanation of the universe, we should regard them as partners in its discovery. They would meet on the ground of common modesty. Each of the three, by frankly admitting that the fact before it was not the whole of the universe but the merest fragment of it, would be in a mood to combine its efforts with the other two for enlarging that fragment into something more significant and more satisfactory. ... Science, philosophy, and religion are not, then, rivals for the hegemony of the spiritual world, but coadjutors in a common enterprise, and the more there is of any one of them, the higher the part the others can play."[4]

Such is the trend toward a genuine reconciliation of scientific knowledge and religious experience, which is so hopeful a portent in the liberal religious thought of today. Its vital importance and broad implications should everywhere be taken to heart. When the average man and woman comes to realize that here is a pathway leading to a fuller life, rich with new values alike for the mind and the spirit, "the conflict between science and religion" will no longer be a live issue. Perceiving that the highest manifestations of mind and spirit are not basically opposed, but that, on the contrary, they are capable of harmonious coordination, such conflicts will fade away because we shall see that they are needless and, in the larger sense, unreal.

2. W. L. Sperry, in *The Atlantic Monthly,* June 1924, p. 767.
3. Professor E. G. Conklin, *The Direction of Human Evolution,* p. 244.
4. L. P. Jacks, in *The Atlantic Monthly,* February 1924; pp. 230-232.

THE HOPE OF SCIENCE

Chapter VIII

The Hope of Science

Dynamic change is the keynote of our time. We apparently stand at one of those momentous crises in human history when mankind moves into a new age. Such crucial periods are of supreme importance, because their outcome may determine man's course for many generations—perhaps for many centuries.

Change spells struggle. This is certainly true of today. The outstanding feature of modern life is the vast flood of new knowledge and power that is sweeping in upon the established traditions of the past. Venerated beliefs, cherished customs, and settled details of everyday life portentously alter before our startled eyes. Across our modern world strange winds are blowing. Perplexed by new problems and aware of coming trials, we confront a formidable future.

Yet, to those aware of the realities of our situation, apprehension is shot through with an expectant thrill. To them, the numberless marvels of technical invention that have transformed our material world are merely the outward expressions of a method and spirit capable of bearing us onward to a future almost beyond our fondest dreams. Such persons are undaunted by unprecedented perils because these are linked with unprecedented opportunities. This adventurous, pioneering note increasingly characterizes the thought of our time. More and more we are coming to realize that Science is, indeed, the *Zeitgeist*, the Spirit of the Age. And, comprehending as we begin to do the broader implications of science, we realize that therein lies our hope.

What is needed is to drive home this understanding of the practical application of science to everyday life. When the average man and woman begin to see that the same attitude and method that the scientist employs in his laboratory and the inventor uses in his workshop will *pay big dividends* if applied to the general run of public and private affairs, we shall see a world wherein knowledge and clear thinking will prevail over ignorance and prejudice much more often than is now the case.

And one of the best proofs that the trend of the times lies in that direction is the way large sections of the general population are plunged, willy-nilly, into a scientific environment from which they cannot fail to absorb more or less of the scientific attitude. The skilled mechanic and up-to-date farmer, for instance, know that their bread and butter depends on exact knowledge and an alert open-mindedness to the latest improvements. They know that to act on old-fashioned ideas and out-of-date methods spells stagnation or failure. They may care little for anything scientific that cannot be turned at once into dollars and cents; nevertheless, they are unconsciously acquiring the scientific viewpoint through their habitual use of scientific knowledge and methods. Slowly but surely their attitude is being transformed, and sooner or later they will come to see that what pays in their own business will also pay in political and social matters and in their general outlook on life. Thus are being laid the foundations of a more rational and enlightened public opinion, and this process (unless interrupted by some political or social cataclysm that might wreck our civilization) should ultimately be a mighty influence for progress.

How, then, can the process best be hastened so that progress may be furthered and possible disasters averted? Our unhesitating answer is: By a general attitude of mind rather than by any specific program of action! The great pitfall of reformers is that they try to evolve high-sounding "principles" or construct attractive-looking programs, which, they confidently assert, will land us in a perfect and permanent Utopia. Our reformers seldom stop to ask themselves how they got such brilliant ideas. If they would only analyze their own minds, they would find that these ideas are, almost always, mere "rationalizations" of emotional beliefs and prejudices. But that is precisely the vice that they condemn in their "reactionary" opponents! Both radicals and "stand-patters" are thus tarred with the same brush, since both either ignore or reject the scientific method and attitude.

Here is where the spirit of science stands forth as a new element in human history. Hitherto man has been guided almost exclusively by either an uncritical belief in traditional authority or an equally uncritical enthusiasm for emotional change—both in equal disregard of *facts*. Thus, Man has not only known comparatively little but has tended to misinterpret the little that he knew. On slender fact bases he has reared elaborate theories, finely spun from his logic and imagination, and he has then crystallized these theories into *beliefs* so dogmatic and intolerant that they have blinded his vision and closed his mind. Society has thus continually ossified, and the few free souls who sought truth with single-hearted devotion have been crushed by the prejudice and passion aroused at the mere thought of examining matters that had become cherished faiths. Human progress has thus far been like a series of lava flows, at first moving with hot haste, yet soon cooling into a rigidity that might be broken or worn down but that could not be kept long in motion. Now, for the first time, we have in the scientific spirit a force capable of maintaining steady and consistent social progress. Its passion for truth can keep us going, while its insistence on proving and testing each step of the way can keep us going *right*. A society genuinely imbued with the scientific spirit and using scientific methods could neither ossify nor run wild. It would thereby avoid both reaction and revolution—those twin ills that have so afflicted mankind.

If our leaders would only appreciate the transcendent importance of this and would concentrate their efforts on the task of *freeing and opening the mind,* instead of consuming their energies in rival programs and propaganda, the level of our political and social thinking would be raised to a higher plane, and the problems in which they are most interested would thereby be much nearer solution.

As things now stand, our reformers are literally putting the cart before the horse. What is the good of hatching new schemes if those schemes are rooted in uncritical reliance on authority or emotional rationalizing—those old, old stumbling blocks in man's upward path? Survey the various programs for dealing with current problems: whether they be labeled "conservative," "liberal," "radical," or whatnot, they are all more or less dogmatic, uncritical, and disregardful of reality. Consciously or unconsciously, they all tend to justify themselves by abstract "principles" with which the facts of the case must somehow or other be squared—or conveniently ignored.

That, of course, is the complete reversal of the scientific method, which begins with a careful search for the facts, faces them frankly, and refuses to dogmatize, realizing that the "truth" of today may be superseded tomorrow by fresh evidence or by a sounder interpretation of existing data. Can there be a greater contrast between the scientific attitude of the "working hypothesis" and the open mind and, on the other hand, the traditional leaning on "established principles," which may hardly be questioned without a sense of guilt and fear? That, and the equally irrational tendency to swing to the opposite pole of emotional change, are the immemorial handicaps to human progress. Can there be an ideal more challenging and inspiring to forward-looking men and women than the task of enlightening their fellows as to how their minds may be freed from these ancient errors? Surely a rational, dispassionate attitude is the necessary prerequisite to the sound consideration of any problem.

This does not imply an underestimate of current difficulties. In the preceding pages, we have surveyed the problems of modern life and have discovered how grave they are and how urgently they demand our best efforts toward their solution. However, they can be solved only by wise and careful handling. Truly constructive action can proceed only from creative thought. Unwise and impulsive action merely makes matters worse. Today, perhaps as never before, "a little knowledge is a dangerous thing." For example, a sixteenth-century theological attitude, cleverly using modern psychology and up-to-date propaganda methods, might conceivably throw back the religious life of the masses hundreds of years, while Stone Age political morality, armed with bomb-dropping airplanes and poison gas, could easily destroy civilization.

The supreme difficulty of modern life is, therefore, not the obvious ills that we talk so much about but our attitude toward them. The only remedy lies in our recognition of the basic fact that our new scientific world requires a correspondingly new outlook, which the scientific spirit is alone capable of engendering. By science we can so reconstruct our general philosophy of life that we can surely and safely make use of the new tools and methods that science has placed in our hands.

That is the great point to remember because it is the most hopeful element in the situation.

Though the mighty forces of nature we have evoked may loom like Frankenstein monsters threatening to destroy us, we must not forget that the same knowledge that summoned them to our service can supply the methods for their control. Our new knowledge has come upon us so suddenly and in so haphazard a fashion that we are not yet making proper use of our new powers. In order to do so, we need to see clearly and think straight. And it is the scientific attitude and method that, by freeing us from ignorance and false beliefs, can open our eyes and liberate our minds.

Then, and only then, will we be in a mental attitude capable of dealing constructively with the complex problems of modern civilization. Then, and then only, can we make free use of our marvelous opportunities, coordinating our new knowledge and powers toward rational, progressive ends. When that day comes, we shall find that our present delusions about political and social cure-alls, "solving words," and other magical incantations will have happily abated, because we will be looking at *realities* instead of shams.

Let us cast our vision into the future in order to glimpse what a society genuinely inspired by the scientific ideal would imply. In the first place, its ideal would be *realism.* That would be a striking new development, because no human society has ever yet adopted a truly realistic attitude. What has been misnamed "realism" has been a narrow materialism denying all not evident to the senses as vain or non-existent; whereas true realism, as expressed in modern science, recognizes the most intangible as well as the most palpable; demands reality, yet understands that reality is infinitely varied; seeks truth but knows that truth manifests itself in countless ways. The genuine realist (best exemplified among modern scientists) has a bold mind—and a humble heart. He is neither cast down by failure nor puffed up by success. His sense of balance and proportion is never obscured. Inspired by a quenchless passion for truth, he combines knowledge and idealism in the synthesis of a higher *wisdom* as his guide in his exalted quest.

Here, at last, we have a truly *progressive* attitude. The basic flaw in nearly all past ideals has been that they have been essentially *static.* In other words, the ultimate goal has been Utopia—a state of permanent and changeless perfection, no matter how great the changes necessary to its attainment. We now know that Utopia is a vain delusion that never has existed and never can exist. However strongly it may appeal to our emotional cravings, *it simply is not life.* Science discloses the fact that all life, from protoplasm to man, is, in the last analysis, a *rhythmic tension* requiring perpetual adaptations, and that the cessation of this essentially changeful activity spells *death.* We may populate Utopia with imaginary beings, but we could never live there ourselves!

Now note the momentous consequences involved in the abandonment of the Utopian dream for a truly realistic philosophy. It would imply nothing less than the dethronement of emotion by intelligence in our general attitude and would profoundly affect all our actions. Utopia, whether cast backward into a legendary Golden Age to be recovered or projected forwards toward some millennium to be attained, is essentially emotional in character and therefore causes a correspondingly emotional trend in attitude and conduct. Men tend to swing between extremes of irrational optimism and pessimism as they are either inspired by vain hopes or cast down by the shattering of vain illusions.

This largely explains the unhappy cycles of revolution and reaction that have so afflicted mankind. A truly realistic society would tolerate neither but would advance steadily along the path of progress. Realizing that perfection is impossible, it would enthrone in its stead the ideal of a constant *perfecting* toward illimitably higher destinies. Nature teaches us that not only progress but even continued existence depends upon continual effort. Automatic progress is not implied in the cosmic scheme, and on the whole, it is easier to go backward than forward. Society, like every form of life, must remain plastic, since fixity means the end of development and usually implies the beginning of degeneration, while rigidity spells the approach of death.

Furthermore, the full significance of social plasticity should be clearly understood. We must realize that, in everyday life as in the laboratory, the same scientific methods of dispassionate observation, controlled experiment, and open-minded deduction are needed for the best results. If the scientific method is followed, society simply cannot go fatally wrong, because, though mistakes will undoubtedly be made, they will assuredly be discovered and corrected in time. How different this is from the traditional plan of evolving some absolute "principle" and then following it even in the teeth of unfavorable experience, bound by its authoritative command to proceed toward some theoretical goal! Unhappy pathway to Avernus, descending ever deeper into the emotional quicksand from which men seek escape by those vain flounderings termed Revolution and Reaction!

One of the most valuable results of a widespread adoption of the scientific outlook would be a genuine understanding of the respective roles played by the forces making for social stability and for social change. Our emotional attitude toward life tends to make us visualize these forces as opposed and, in a sense, hostile to one another. We build up in our minds formidable antitheses with combative labels such as "Reaction versus Revolution," "Order versus Anarchy," "Reason versus Faith," and many more besides.

Yet these are mere emotional shibboleths, quite out of touch with reality. As a matter of fact, stability and change are both vital to social life and should be regarded as interacting, not opposed. Stable continuity is as necessary to society as it is to our own bodies. If the bonds of social custom and restraint were loosed and the group should resolve itself into an anarchic mass of individuals, each of them going his independent and unpredictable way, the effect would be the same as though the law of gravitation had suddenly been suspended for our bodies—which would instantly dissolve into chaos. But this does not mean that, within the bounds of social stability, adaptive change is not as necessary as it is in our physical selves if health and even life are to be maintained. It is only when some people emotionally exalt stability and others similarly exalt change that we get the mutually destructive concepts of reaction and revolution.

The scientific attitude can free us from these melancholy dilemmas by enabling us to make a just estimate of the situation. The scientist realizes that one of conservatism's most useful functions is its sifting and testing of novel ideas and methods. It puts the burden of proof upon the newcomer and forces a trying-out of theoretical claims. The struggle for existence among ideas is society's laboratory, and the whole social experimental method is dependent upon the factor of human conservatism, which causes and renders intense the ordeal through which new ideas must pass.

On the other hand, the scientific attitude hastens the abandonment of old established ideas that have outlived their usefulness and have become handicaps or evils. Refusing to regard any custom or institution as sacrosanct and eternal, science looks at all parts of the social structure with objective detachment, viewing them as scientific phenomena to be observed and appraised without prejudice. In this way rational change is made possible without falling into the emotional error of desiring change for the sake of mere novelty or for the subjective satisfaction that many emotional natures derive from the process of change regardless of ultimate consequences.

Such is the vista of sustained, balanced progress open to a society genuinely inspired by the scientific spirit and employing scientific methods. How to infuse this spirit and these methods into our existing social order, how to direct the idealistic vigor and intellectual energy of our age into truly constructive channels, and how to weld the clashing, uncoordinated elements of Old and New into a harmonious vital synthesis is the supreme problem of modern life, on whose solution the future of our civilization depends.

SCIENTIFIC HUMANISM

Chapter IX

Scientific Humanism

A new age is at hand. Its dawning lights the sky. What is that new age to be? Will it witness a fresh flowering of the human spirit?—A great advance along the path of progress? Or will the rose tints of dawn be overcast by clouds heralding a day of storm and stress and closing in the night of another Dark Age?

Questions like this are being put and pondered everywhere by forward-looking minds. They are no idle queries, because upon their answer depends man's destiny for centuries to come. And what the answer shall be depends largely on ourselves. It is upon us living men and women, born as we are into this crucial transition time, that there is placed the high privilege and the high responsibility of determining the trend of the future.

Faced with such exalted opportunities, we can neglect no source of guidance to aid us in our task. And history furnishes us one clue of peculiar significance. This is the period known as the Renaissance (literally "The New Birth"), when the era now obviously passing was born. Somewhat over four hundred years ago, on the threshold of the sixteenth century, our ancestors confronted a crisis of transition strangely like our own. How they met that crisis and how far they succeeded and wherein they failed may serve alike as inspiration and as warning for us today.

Let us cast our mind's eye back across the centuries to the momentous period that gave birth to what we are wont to term "Modern Times." The thousand years of intellectual and spiritual immobility known as the Middle Ages was drawing to a close. The mind and soul of the race, so long fixed in certain basic molds of thought and feeling, were breaking with tradition and questing eagerly along new paths. It was a wonderful time. The touch of Spring was in the air. Men felt that the world had become young again and that a true Golden Age lay on the horizon.

Now what produced this marvelous change? Primarily, the impact of new knowledge and new ideals. The fact of the matter was that the men of the Renaissance had uncovered a long-buried world—the Classic World of Greece and Rome, which had been entombed in the night of the Dark Ages a thousand years before. The effect of this rediscovery of the Ancient World with its intellectual freedom and courage and warm love of life, so utterly different from the rigid dogmatism and narrow outlook of the Middle Ages, was astounding. The cramping bonds of medievalism were torn asunder, and the liberated mind glimpsed a richer, freer life almost beyond its fondest dreams.

For the moment, all seemed plastic and anything possible. Many ardent spirits caressed the wildest fancies and the most visionary schemes. The wiser minds, however, were not deceived. They saw that the plasticity of their transition epoch might be used with great profit in shaping the future, but they were also aware that the future must be built on stable foundations. Hence, they realized that a revival of the pagan past, as was suggested by some, or the creation of a wholly new Utopia, as was proposed by others, were alike impossible. What the constructive thinkers of the Renaissance had in mind was a genuine assimilation of all that was best and fittest of the legacy of classic antiquity into their Medieval world, resulting in a new order based upon a harmonious synthesis of both elements.

The most interesting point about this plan was, however, not so much the general conception as the way in which it was to be brought about. The method suggested was, in the last analysis, an intellectual and emotional attitude or philosophy of life called "Humanism." The term comes from the words *"litterae humaniores"* ("humane letters") used to denote the Greek and Latin classics, which were prized by the men of the Renaissance because of their "humanizing" effect upon the individual. A "Humanist," therefore, was one who, having envisioned the freedom, beauty, and joyousness of the Ancient World, with its emphasis upon personality and its keen interest in life, was emancipated from the ascetic dogmatism of the Middle Ages, which despised this world and cramped the individual in a rigorous discipline for the next world—the only thing that really mattered.

What could be a more striking contrast than the Humanist and Medieval ideals? Their standards of value were diametrically opposed. The Middle Ages set up a criterion of truth that was absolute and eternal, arbitrarily imposed upon the world and "divinely" revealed through an infallible Church to which complete obedience was required on pain of the direst penalties in this life and everlasting torment in the life to come. However fair the visible world might seem, it was a snare of Satan. The less attention paid to it, the better if Heaven were to be attained by the soul, temporarily encased in its sinful, fleshly garment.

Against this medieval concept Humanism uttered a ringing protest. Acting on the motto of the Greek philosopher Protagoras, "Man is the measure of all things," the Renaissance thinkers formulated a standard that was at once human and relative. Life was conceived as a perpetual process of adjustment of human individuals to the world and to one another. The ideal was of an organic, perfecting harmony rather than a static, changeless perfection. To be sure, the Renaissance Humanists did not proclaim their philosophy so baldly and boldly as I have here stated it. Nevertheless, the implication is clear.

Renaissance Humanism is well expressed in the life and writings of Erasmus. This brilliant Dutch thinker incarnates all that is best in the movement. His outstanding quality is a broad reasonableness. Urbane, reflective, open-minded, and tolerant, Erasmus devoted himself to the cause of sound thinking and common sense as opposed to the forces of ignorance, superstition, and bigotry. Dogmatism and fanaticism he abhorred as the most destructive of evils because he realized that by rousing the passions and closing the mind, they blocked the path to enlightenment and genuine progress. His ideal was a harmonious, Humanistic civilization, combining the genius of antiquity with the Christian spirit and inspired by a true blend of idealism and reason.

Alas! The Golden Age envisioned by Erasmus and his compeers remained a dream. The bright hopes of the Renaissance were blasted by storms of strife that plunged Europe into a welter of wars that raged for generations. The dreary cycle ended with Europe politically divided into chronically hostile states ruled by despotic monarchs and spiritually sundered into a group of dogmatic, intolerant Protestant churches confronting an even more dogmatic and intolerant parent church in which the liberal tendencies latent in Medieval Catholicism had been eliminated by a rigid and uncompromising Romanism.

In such an atmosphere of intolerance and fierce partisanship, Humanism simply could not live. Tolerant, reasonable minds like Erasmus were hated and vilified by all parties as trimmers and traitors. "He who is not with us is against us!" was the general slogan, and men were compelled to take sides or court almost certain martyrdom. The whole Humanistic attitude became intolerable to an intolerant age. Keen-sighted zealots, Catholic and Protestant alike, came to see in Humanism a common enemy. And they were right, because the new order envisioned by the Renaissance prophets implied mental and spiritual changes far greater and deeper than those imposed by the Protestant Reformers, and hence much more profoundly disquieting to the champions of the Catholic reaction.

Accordingly, in both Catholic and Protestant lands, the Humanistic spirit was systematically weeded out. Formal persecution was impracticable because Humanism, by its very nature, was not a dogmatic creed or organized sect, and there was therefore nothing tangible to persecute. Yet the campaign against Humanism went on nonetheless, by indirect methods. This is well shown by the way in which Humanism's wellsprings, the Greek and Latin classics, were tampered with and perverted.

The men of the Renaissance had prized the masterpieces of Greece and Rome not only for their intrinsic beauty but even more for their "humanizing" effect—in other words, for their power to open and free the mind. Once a man had really grasped the meaning of that ancient world with its intellectual freedom and its spiritual virility, the Medieval bonds galled past endurance and were instinctively cast aside. The Renaissance leaders thus used the classics as their chief weapon in their effort to break up the rigid, dogmatic Medieval system.

Such was the Humanists' method—a method employed so successfully that their dogmatic opponents were faced with the logical alternative of either abolishing the classics or nullifying their liberalizing effects. The printing press having rendered physical destruction of the classics impossible, the latter course was adopted. The classics were still honored but were so taught that their effect upon the mind and spirit was wholly altered. Employed mainly as a means of discipline and as a badge of intellectual snobbery, "The Humanities" soon ceased to have their original Renaissance meaning and, in fact, became actually invested with a superstitious veneration as containing *in themselves* a unique cultural significance. What a perversion of meaning and inversion of values!

Despite its bright promise, therefore, Renaissance Humanism was, in the deepest sense, a failure. The harmonious Humanistic civilization envisioned by Erasmus and the other Renaissance leaders never came into being, and the Humanistic movement itself was nullified by the growth of religious and political dogmatisms, which rent the Western world for generations and which still vex us today. Of course, the legacy of Greece and Rome was taken over after a fashion, but it was taken over only partially and asymmetrically. Can anyone seriously deny that the classic influences, which so quickened and vitalized the men of the Renaissance, might, under happier auspices, have remained a stimulus of the first magnitude in our civilization instead of subsiding into the cultural formalism that renders them a diminishing factor in the world of living realities?

Such is the story of the great awakening of the human mind and spirit, which took place four centuries ago. It is a tragic tale with a warning so clear to our own day and generation that we disregard it at our peril.

What striking parallels does due reflection disclose! We, like the men of the Renaissance, live in a time of profound change and are faced by pressing problems of transition. As the rigid edifice of the Middle Ages was at last shaken by the dynamic impact of new knowledge and ideals culminating in the rediscovery of the classic past, so our existing (and mainly traditional) order is being profoundly disturbed by the onrush of a flood of new knowledge, methods, and ideas, which we broadly term modern science. Somehow or other these new elements must be genuinely assimilated into the fabric of modern life; somehow or other Old and New must be welded into a true synthesis if disharmonies and conflicts are to be avoided, which may wreck our civilization.

Now, are we not facing a problem strikingly like that which confronted our ancestors four centuries ago? Of course, the elements to be assimilated are quite different, but the process of assimilation is basically the same. Let us beware lest modern science be incorporated into our idealistic and cultural scheme as partially and asymmetrically as was the classic legacy into the post-medieval epoch—with similar unhappy results!

Why did the men of the Renaissance fail to attain their larger goal? Why did Humanistic civilization never materialize? Because the movement operated on too slender a basis. The Renaissance was virtually confined to a small, educated minority, almost out of effective touch with the bulk of the population. Indeed, concerned as the Renaissance was mainly with the classics and conducted as it was almost entirely in Latin and Greek, it had no practical way of enlisting the interest of the masses. Therefore, the Renaissance movement, though brilliant and sound, lacked deep foundations. Like a house built upon the sands, it could not stand the buffetings of storm and stress.

What a pitiful tragedy was its failure! The leaders of the Renaissance were men of sanity and vision. In Humanism they had an ideal and method capable of producing the happiest results. Erasmus and his fellow Humanists saw, as few men of any age have seen, that the path of true progress lies not through overthrowing an old by a new dogmatism but by freeing the mind and spirit, thereby opening them to the influence of reason and thus ridding men not merely of a particular dogmatism but of the dogmatic *habit* as well.

The Humanists also realized the gathering forces of passion and unreason that menaced the realization of their golden dreams. They appreciated the fact (so well analyzed by modern psychology) that in times of change the shock of ideas is peculiarly intense and that the tendency of human nature under such circumstances is to develop partisanship, intolerance, and fanaticism in which even the wisest and coolest men may lose their vision and be swept away by the emotional tide. To prevent the flood of passion from reaching that danger point, Erasmus fought his gallant fight, denouncing Protestant and Catholic zealots with equal vehemence and pleading for reasonableness, tolerance, and moderation. In vain! The voice of reason was lost in the roar of the emotional flood, and the banner of Humanism disappeared in the raging waves of sectarian strife. Generations of war and misery were to elapse before the hard school of suffering was to teach Europe a part of the lessons that the Humanists had proclaimed.

Apply this to our own day. Once again, a wealth of new knowledge has quickened our minds, and a new idealism has appeared to elevate the human spirit. Taken in its larger sense, modern science can give life new values and new meanings, while the application of scientific knowledge and methods to current problems is capable of raising the level of civilization to a higher plane. The age that is coming will be an age of true enlightenment and progress if we succeed in really assimilating the vast extensions of knowledge and power that we have amassed into our present idealistic and cultural scheme. And the way to do it is the way already charted by the men of the Renaissance; that is to say, by an attitude of mind and spirit, which we may term *Scientific Humanism.* A new Renaissance is palpably "in the air." Its essential elements are present, ready to hand. We need only an appreciation of the significance of our new knowledge and power and due understanding of their use to ensure the appearance of this new Renaissance and its flowering in a new Humanism that will endure.

But will it come? That is the anxious query that must spring to the lips of every thoughtful observer of our time. In preceding chapters, we have surveyed the problems of our day and have discerned the perils lurking beneath the surface of modern life. We have noted its grave disharmonies and maladjustments, together with the ominous increase of emotionalism, which is perhaps the most disquieting symptom of the times. Today, as in the Renaissance four centuries ago, the forces of ignorance, passion, and unreason are astir; now, as then, reaction and fanaticism are alarmed and powerful. Will those primitive, volcanic forces, always latent in even the most cultured communities, burst forth to overwhelm the coming Renaissance and perhaps shatter the whole fabric of our civilization?

That our civilization is in danger no thoughtful man can doubt; yet that the danger can be averted few will deny. Certainly there will be no catastrophe if we will but follow the light with which science has illumined our path. To point out that light to the masses of mankind and to explain to them the things it discloses is the saving task of the intelligent minority whom the scientific spirit has infused with the new Humanism.

Herein lies our chief advantage as contrasted with the Renaissance of four centuries ago. The Renaissance Humanists were but a handful, quite out of touch with the mass of their fellow men. Today, on the contrary, not only is the enlightened minority very much larger, but the new scientific knowledge directly touches the everyday lives of the majority and awakens practical interest. The common man may not yet understand, but he can be encouraged to understand at least the broad significance of what a Humanistic age, deep-rooted in the soil of scientific knowledge and flowering in the genial air of enlightened idealism, would mean *to him*. When he has grasped the possibility of a richer, fuller, happier life in such an age, the common man will hearken to the voice of reason rather than to the *siren song* of passionate emotion.

Such, at least, is the hope and the task of all those whom the scientific spirit has touched and transformed. In the consciousness of their exalted opportunity, let them raise aloft the banner of Scientific Humanism—a standard of enlightened goodwill to which the wise and righteous may repair.

Index

www.ingramcontent.com/pod-product-compliance
Lightning Source LLC
LaVergne TN
LVHW022309200926
848481LV00004B/1253
* 9 7 8 1 5 9 3 6 4 0 7 0 5 *